Innovation and Regional Development:
Strategies, Instruments and Policy Coordination

Innovation and Regional Development

Strategies, Instruments and Policy Coordination

Proceedings of the Fifth International Conference on Innovation and Regional Development held in Berlin, December 1–2, 1988

Edited by
Hans-Jürgen Ewers
Jürgen Allesch

W DE G

Walter de Gruyter · Berlin · New York 1990

Editors

Hans-Jürgen Ewers
Institut für Verkehrswissenschaft an der Westfälischen Wilhelms-Universität, Münster

Jürgen Allesch
Technologie-Vermittlungs-Agentur Berlin e.V.

Library of Congress Cataloging-in-Publication Data

Innovation and regional development : strategies, instruments and policy coordination ; proceedings of the 5. International Conference on Innovation and Regional Development, held in Berlin, December 1–2, 1988 / ed. by Hans-Jürgen Ewers ; Jürgen Allesch. – Berlin ; New York : de Gruyter, 1990
ISBN 3-11-012261-8
NE: Ewers, Hans-Jürgen [Hrsg.]; International Conference on Innovation and Regional Development <05, 1988, Berlin, West>

Library of Congress Cataloging-in-Publication Data

International Conference on Innovation and Regional Development (5th : 1988 : Berlin, Germany)
Innovation and regional development : strategies, instruments, and regional policy coordination : proceedings of the Fifth International Conference on Innovation and Regional Development held in Berlin, December 1–2, 1988 / edited by Hans-Jürgen Ewers, Jürgen Allesch.
X, 346 p. 17 × 24 cm.
Includes bibliographical references.
ISBN 3-11-012261-8 (Walter de Gruyter, Berlin). – ISBN (invalid) 0-89925-620-5 (Walter de Gruyter, N.Y. : acid-Free paper)
1. Regional planning – Congresses. 2. Europe – Economic conditions – 1945 – Regional disparities – Congresses. 3. Technological innovations – Economic aspects – Congresses. 4. Technological innovations – Social aspects – Congresses.
I. Ewers, Hans-Jürgen. II. Allesch, Jürgen. III. Title.
HT391.I472 1988
338.9–dc20 90.3870
CIP

⊗ Printed on acid-free paper

Layout: B. Busch, Berlin. – Printing: WB-Druck, Rieden. – Binding: H. Stein, Berlin.

Preface

This volume features the papers presented at the Fifth International Conference on Innovation and Regional Development which was held in Berlin (West) in December 1988 and was attended by a wide audience from many European countries, Canada and the USA. It continued the successful series of annual conferences held on the subject of Science Parks and Innovation Centres started in Berlin in 1984.

We are grateful to many individuals and organisations who helped to make the Fifth International Conference on Innovation and Regional Development the success it was: the European Business and Innovation Centre (EBN) and the European Association for the Transfer of Technology, Innovation and Industrial Information (TII) who were co-sponsors of the conference; the Commission of the European Communities, Directorate XVI (Regional Policy) and the Senate of Berlin for their endorsement and financial support; Wolfgang Gessner, Karin Esders-Angermund and David Weissert for the organization of the conference; and David Weissert for taking charge of the tasks involved in bringing this book to publication.

Hans-Jürgen Ewers - Technical University of Berlin

Jürgen Allesch - Technology-Transfer-Agency Berlin (TVA)

Contents

3 Regional Development Potentials and Strategies for Strengthening Regional Innovation Capacities

4 Instruments of Innovation Oriented Regional Policies

Welcoming Address

Elmar Pieroth

The International Conference for Innovation and Regional Development is taking place in Berlin for the fifth time. The presence of 240 experts from 21 countries underscores the significance that the issues of regional policy have attained in the mean time. One does not have to itemize the issues to realize that regional development is more and more looked upon as an independent undertaking. We believe that in the case of Berlin this undertaking requires its own special treatment. Berlin is a region on its own and, as such, has its own opportunities and problems. For almost two decades these problems led to an annual loss of about 8,300 jobs in industry and to an annual out-migration of nearly 20,000 Germans. In recent years, however, this tendency is being reversed. Berlin, once again, is a growing metropolis with immigration, decreasing death rates and proportion of retirees and a place where jobs are once again being created. Furthermore, in a region such as Berlin, more than anywhere else, it is vital to determine how the endogenous resources can best be utilized, how more can be achieved by the region's own effords, on the one hand and with outside help on the other.We believe that Berlin is the suitable location for a series of meetings to discuss the effectiveness of strategies and instruments of innovation-oriented regional policy, to search for ways for the institutional coordination of innovation measures, and to examine to what extend the concepts of regional policy are transfered. Berlin is certainly in the forefront of innovation-oriented economic policy and, therefore, a model where the issues of this conference can be studied in practice. After all, what we want is an improvement of conditions for all people in all regions.The progress of science, which has always been connected with achievements in innovation, and the development of science based civilization, have brought a quality to people's everyday lives that would not otherwise have been possible. The time, freedom, choice, tolerance, mobility and communication thus won make possible an increase of mutual understanding also internationally. This can be seen as a gain for the individual in the humanization of labour. Included in this idea is the delocalization of jobs from companies. The principle of centralization and therefore concentration of employees, that established itself with industrialization, may be replaced in future by a decentralization of jobs. The fusion of electronics and telecommunications may allow for the interaction of people, even when they are not working at the same place or the same time giving working at home a new quality.We are planning to develop the results of this conference into scenarios for a European economic policy. Therefore, we are pleased, that many visitors from the southern and eastern parts of Europe and also from overseas are participating. We are pleased as well that visitors from the German Democratic Republic and from Hungary are taking part. Cooperation between the economic blocks is a task which we in Berlin have always especially emphasized. The Western Europe of 1992, without borders is not intended as a fortress locking its gates to other economic regions. What applies to the relationship to North America and to Asia applies especially to the relationship to our COMECON neighbour states. Our special interest in Berlin and, therefore, our special obligation is to foster an interaction of the two economic systems. I am glad that this aspect is also included in your work and thus also for this reason sincerely welcome you in Berlin.

Introduction

I.

The 1980s can be viewed as a period of major changes in regional development in Western Europe. The general slowdown of economic growth that had characterized the 1970s eventually came to an end during the second half of the decade, with considerable impact on regional patterns of growth. Agglomerations, having suffered the heaviest losses of employment during the recessions of 1974/75 and 1981/82, gradually experienced a comeback while "Take off regions", which had gained from the first wave of industrial restructuring increasingly became confronted by the limitations of their late industrialization. As a result, the regions of Europe today exhibit a much more complex and diverse picture than was the case in the 1960s when classical regional policies were first implemented. These policies attempted merely to redirect transferable investment in industry from agglomerations to disadvantaged regions, in the hope of inducing regional mulitiplier effects. Their main instruments were financial grants and subsidies to investment. However, this approach to regional development clearly lost ground when, for a number of reasons, the envisioned potential for interregionally mobile investment vanished during the 1970s.

Competition for the decreasing number of transferable investment projets severed, causing unwelcome picking-up effects and a bidding up of levels of grants and subsidies, without inducing additional investments. Furthermore, the long run effectiveness of investment-oriented policies of regional development began to be viewed with greater skepticism. In many cases, the structural weaknesses of regions proved to be not curable by industrial investment alone. "Cathedrals in the desert" were built bearing hardly any relationship to the existing structures of the regions in which they were located. Regional multiplier effects, by and large, did not work.

The fairly substantial growth that some of the disadvantaged regions of the 1960s had enjoyed relative to the declining agglomerations, could mainly be accounted for by labour and pollution-intensive industries with, however, decreasing value-added content due to increasing international competition. Meanwhile, the technology-led growth of the 1980s turned the tide in favour of those agglomerations which had undergone a successful process of industrial restructuring. The dynamics for growth were best in regions with a strong research and development infrastructure where micro-electronics and other high-tech related industries could flourish. Innovation, both of products and processes became the single most important factor of economic success.

As a response to the increasing uneveness of regional development and the changed economic situation, the emphasis shifted away from redirecting the growth potential to clearly identified disadvantaged regions simply by supplying financial incentives. Current policies attempt to develop new growth potential within regions, taking as a starting point a region's endogenous resources of businesses and skills. The objective is to develop the capabilities of small and medium-sized enterprises (SMEs) from within a region to invent and innovate, and thus to develop new products for new markets. The promotion of investment is being replaced by the promotion of innovation.

This change of paradigm in regional policy has been accompanied by a shift of policy levels. The decentralized nature of the new innovation-oriented endogenous policies means that regional policy levels have gained ground relative to national governments which are at the same time losing to the European Commission jurisdiction over policy. Thus, national governments find their role being increasingly reduced to coordination of policies initiated within the regions themselves and in direct cooperation with the European Commission.

II.

The impact of these innovation-oriented endogenous policies on regional development and the problems arising with their implementation are assessed in this volume. In **Part one** current issues of regional development are introduced from a practical point of view. *Hahn* gives a brief overview of the various activities arising as a response of regional policy to the challenge of the "third industrial revolution". Drawing on the experience of the German Federal State of Saarland, he especially stresses the importance of cross-national cooperation. As an entrepreneur and chairman of a development agency *Jones* addresses regional development by calling for effective advisory bodies to support innovative enterprises and generate an entrepreneurial climate within a region. *Mulcahy* reports on the changes in the European Commission's approach to regional policy brought about by the Single European Act. He especially emphasizes the central role that the Regional Fund will play in overcoming regional disparities, outlining the elements of a future "total system approach" to the development of the regions of Europe.

Part two includes two studies of regional disparities in Europe arguing the case for an active regional policy. *Genosko* surveys the regional disparities existing within the EEC and briefly discusses their possible causes. He stresses the impact of existing industrial structures hindering new developments and of general living conditions on a region's development chances. *De Gaudemar* presents a methodology for the analysis of regional disparities. Outlining and interpreting different typologies of regional disparities he concludes that tendencies towards a convergence of regional development in Europe persist, which are, however, partly countered by diverging developments, especially in southern parts of the European Community.

Part three addresses the question of regional development potentials and strategies for strengthening regional innovation capacities. *Townroe* points to the central role of a region's capacity to invent and innovate for its ability to compete. The fostering of "new technology" factors, he argues, is an important objective of regional policy which together with transfer measures should support the creation of a social culture of technological enquiry and corporate adventure. *Davelaar's and Nijkamp's* contribution examines the key role of producer services in the process of spreading new technologies in time and space. Drawing on an extensive case study from The Netherlands, they demonstrate the potential of this sector for rapid growth and the spreading of innovation from agglomerations to peripheral regions. *Malecki* explores the possibilities of fostering the regional environment for innovation and entrepreneurship suggesting that less stress should be placed on the size of a region than on the presence of communication and the contact networks of key individuals. He submits that regional policy should confront the cultural issue involved with the creation of an innovative environment. *Bade* describes the spatial pattern of innovation-relevant services in the Federal Republic of Germany and finds that a high concentration of such services are to be found in agglomerations. This, however, does not seem to hinder a secular disagglomeration trend concerning employment and growth, although the performance of firms with an above-average intensity of producer services is particularly good, both inside and outside agglomerations. *Brugger* examines the analytical content of the term "endogenous development" arguing that two distinct political strategies can follow from it: a strategy of developing and controlling internal relations in a context of interregional exchange or a strategy of qualitative zero growth of a region with little interregional exchange. With both strategies control of internal resources and self-determination are crucial for endogenous development.

In **Part four** instruments of innovation-oriented regional policies are examined. *Allen* studies the key features and trends of recent years with respect to the financial instruments for promotion of regional development in EEC countries. He argues that there has been a shift away from "automatic" instruments to more selective measures as well as away from investment-related to "softer" measures like advice and consultancy. He also identifies increasing attempts to coordinate regional incentive policies with national incentives in related fields. *Kern* discusses the relationship of financial and real transfers to recent developments of regional policy. Although welcoming the shift in paradigm from investment-related financial aid to transfers in kind, he contends that there is a continuning need for traditional measures of promotion. In his short contribution, *Hull* points to the key role of information and consultancy services provided by local and regional economic promotion agencies. He stresses the importance of maintaining a budgetary commitment to secure the funding of such agencies for a period sufficient to make an impact on regional development. *Hjern* raises the issue of regional qualifications structures as a precondition for successful regional development. Drawing on the Swedish experience he calls for educational and training policies to redress regional disparities in qualifications. *Peters* presents the case for strong co-operative relationsships between the universities, industry and regional government to mobilize a region's capacity to innovate, taking the cases of Cambridge (UK) Warwick (UK) and Austin (Texas) as examples.

Part five examines indigenous development measures required in regions with differing growth dynamics. *Camagni and Rabellotti* argue in their contribution the case for advanced technology policies in developing regions. They point to the central role of innovaton and technical progress in the regions' competitive position, rendering it impossible for single regions to lag behind in technology. As spontaneous developments show a centripetal bias against peripheral regions, regional policy must adopt strategies aimed at balancing the arising disparities. The reasons for the long-standing economic backwardness of Ireland, that can not be explained by its geographical peripherality, are explored by *Sweeney* . He submits that the causes of decline are to be found in an amalgam of historical, social and political factors that sum to a lack of technical and entrepreneurial culture hindering the indigenous potential for growth. *Gabriel* compares three success stories of regional development: Baden Württemberg (FRG), Emilia Romagna (Italy) and Massachusetts (USA). Although each of these regions pursued different industrial strategies, a common factor of success can be identified in the ability of each to create formal and informal institutional frameworks, which, geared to the specific needs of the respective region, enabled them to fully utilize their indigenous potential. *Allesch* calls for a holistic approach to regional policy in agglomerations. Drawing on the example of Berlin, he argues that an endogenous strategy should concentrate on the provision of real transfers, i.e. information transfer, technology transfer and personnel transfer.

Part six is a summing up, with general conclusions regarding the outlook on the future of regional policy in Europe. The evidence presented on regional development trends, the forces behind those trends, and the kind as well as the policy level of revitalization strategies likely to be involved are summarized by *Ewers*. He presents "ten commandments" of endogenous innovation-oriented strategies and refers to some unsettled issues. *Hirche* emphasizes that regional policies have to be based within the regions themselves. To be successful, economic policies must be supplemented by appropriate measures affecting education, the culture, and the natural environment. But most important, structural change and innovation must not be hindered by attempts to keep obsolet structures alive. A successful regional policy has to be innovation-oriented and at the same time it must a have an eye on the actual living conditions of the region as a whole.

Jürgen Allesch

Hans-Jürgen Ewers

1 Regional Development: A Practical Point of View

1.1 Innovation and Regional Development in the EEC

Ottokar Hahn

The European single market to be realized after 1992 will bring about considerable changes and will have a strong influence on regional development, not only in the member states of the European Community. It also has the goal to strengthen the competitiveness of the European economy in relation to Japan and the USA. This becomes especially clear when one considers that the EEC countries' share of the international trade in electronic equipment, office equipment, computers and chemical products fell from 35% to 26% between the years 1963 and 1985, while Japan's share rose from 6% to 25% in the same period, and this, even though the member states of the European Community spent twice as much on research and new developments in these fields as their competitors from the Far East. How can the enormous funds that are used for the development of new products and industries as well as for new production methods be used most efficiently (i.e. goal-directed and co-ordinated)? How should regional development be supported through innovation? Which strategies, instruments and forms of organization already exist and what has still to be developed?

As a minister of the Saarland I believe I should share some of my experiences in this field, with regard first, to the EEC as a promoter of "innovation and regional development" , second, regional promotion of economic development in a member state, using the example of the Federal Republic of Germany and the Saarland, and third, cross-jurisdictional co-operation in the field of promotion of original innovation, using the example of Saar-Lor-Lux.

1 Experiences with the EEC as a Promoter of "Innovation and Regional Development"

For many years now worldwide development has been marked by economic and technological progress that has been termed the "third industrial revolution". Mostly this refers to the fields of microelectronics, computer technology and telecommunications. In the future these developments will have a great influence not only on the economy, but also increasingly on the life of each and every citizen, involving even leisure-time activities. This development, which will determine the shape of the coming years and decades, will certainly increase the competition between Europe, the United States and Japan in these fields.

The EEC recognizes this problem and is increasingly trying to strengthen Europe's competitiveness in the international market by putting into effect the European single market by the end of 1992 and thereby creating the world's largest market area with 320 million inhabitants and the ensuing growth effects. The unification of industry standards is undertaken to facilitate the cost-saving advantages of economies of scale. Furthermore, the possibilities of promotion of the Community institutions have been substantially extended with Articles 130f to 130q of the United Europe clauses of the EEC Treaty that came into effect with the Single European Act in July 1987. After difficult negotiations in the Council of Ministers, the Community institutions now have 6.3 billion ECU, that is about 13 billion DM, in research funds at their disposal for the period 1987-1991.

1.1 EEC Programmes for the Promotion of Research and Development: ESPRIT, RACE and BRITE

ESPRIT is short for the European strategic programme for research in information technology. In the second phase of the ESPRIT programme (1987-1991) a special emphasis is being placed on precompetitive research and development activities in the three strategic fields microelectronics and peripheral technology (including the development of highly integrated circuits and integrated high speed circuits), information processing systems and information technology application (e.g. computer-integrated production). Furthermore, as a new element, basic research is to be explicitly related to the fields of microelectronics, computer science, artificial intelligence and cognitive science.

RACE is the programme for research in advanced communication technology for Europe. RACE aims for the installation of a Europe-wide telecommunication infrastructure that is to be available by the end of the 1990s. The possibilities for the use of an integrated broad band network are supposed to be expanded, whereby the improvement of basic and supportive technologies - especially multi-service technologies - have priority.

BRITE is the programme for basic research in industrial technology in Europe. BRITE promotes projects in the field of precompetitive research and development to facilitate the acceleration of the development of new materials and production technologies. Projects of the innovative fields like laser technology, high-density and compound materials, membrane technology and the use of CAD/CAM form the central components of this programme, that (with great resonance and a high degree of participation) has given important impulses to the development and use of new technologies and production processes.

The call for tender for ESPRIT with a total promotion sum of 1.2 billion DM ran this summer. Requests for a total sum of 10 billion DM were received, eight times the monies available. This shows that a considerable need for the support of research and development exists and that the EEC has correctly estimated the need. On the other hand it is remarkable that there is such a great potential for this international challenge in the businesses and research facilities of the European states. Announcements for the programmes BRITE and RACE have been made and a similarly high participation is to be expected.

1.2 Further Programmes for the Promotion of Research and Development

While EEC financing is usually limited to the member states of the European Community, the co-operation in research and development goes beyond these bounds: 19 European states have joined the EUREKA initiative, to accomplish joint projects in the field of new technologies and to commonly finance them beyond the national borders. Meanwhile measures with a volume of 8 billion DM have been decided on for joint projects in information technology, industrial and processing technology, biological technology, medical techniques, new materials, environmental subjects and telecommunications. The research promotion by EUREKA is hereby not to be seen in competition with the EEC promotion, but as a supplement with the goal to combine the existing national instruments of promotion and to use them more efficiently.

One special goal of EUREKA is not only to grant these promotional means to the large-scale enterprises, but also to include small and medium-sized businesses. As an example I would like to mention a development company in the Saarland. A databank is being developed with partners from France, Great Britain and Austria, that is supposed to take the place of the hitherto existing dictionaries and include the principle foundations of grammar and orthography of the three languages German, French and English. This project runs over four years with a volume of 7 million DM, of which 50% is sponsored by

EUREKA. To me this seems to be a good example of the role of the EEC for the promotion of innovation and regional development.

Furthermore, from the regional point of view, other EEC initiatives for the strengthening of the development of small and medium-sized businesses are also to be welcomed. On the one hand, they provide direct promotion of research and development for small and medium-sized businesses, on the other, they aim at the improvement of the flow of information to small and medium-sized businesses. For the first group of measures the technology transfer by the programmes COMETT and SPRINT can serve as example.

COMETT promotes co-operation between universities/colleges and medium-sized businesses in education and advanced training in the field of technology. COMETT is aligned to the education and advanced training of highly qualified specialized personnel in co-operation between universities/colleges and the economy. Its aim is a tighter co-operation of the members of the European Community with regard to technology, research and training, and industrial utilisation. Stays abroad of students, university graduates and university teachers as well as skilled workers involving a limited time at a university or college or in a company of another EEC country, are being sponsored. Training partnerships between universities and the economy are also planned, as are measures of development and advanced training and their testing as well as the obtaining of correspondence course offers with the use of new information and communication technologies. A general exchange of information and experiences rounds off the programme.

The second group of measures - improvement of the flow of information - proceeds from the consideration that small and medium-sized businesses often lack the appropriate equipment with the personnel and organization for receiving the information necessary for research promotion. From this consideration the EEC Commission has so far, in a pilot phase, installed 39 Euro-Guichets as consultation agencies for small and medium-sized businesses all over Europe. In a second phase, the EEC Commission has summoned suggestions for further Euro-Guichets. Additional measures to ease the access to data-banks have been created with the installation of the BC Network Information Service in order to make these EEC information directly accessible to businesses.

The BC Network is the European combination system for business co-operation. The BC Network consists of an EDP system which connects a network of business consultation agencies in all of Europe through data-lines. For this purpose an aid is to be provided for the small and medium-sized businesses that allows them to improve their co-operation in technical, commercial and financial respects or to expand on the European level.

SPRINT is the strategic programme for innovation and technology transfer. In order to support the advance of new technologies and therefore playing an essentiell part in the process of technological renewal and the enhancing of competitiveness of the European economy, the acceleration of the process of innovation and the shortening of delays in technology transfer becomes significant. Innovation is that process through which ideas, coming from research and development or from the discovery of untapped markets, are turned into new or improved, economically viable products, methods or services. The SPRINT programme in connection with the tightly connected Euro-consulting facilities and the enlargement of the consultancy information network (BC NET) have proven suitable for achieving a better integration of small and medium-sized businesses into the economic structure of Europe. Due to the increased speed of the development of the innovation process it will also be significant for the future to further shape the distribution of innovation, the transfer of technology and the creation and distribution of specialized knowledge as completely and effectively as possible. Our experiences show that the EEC watches that the programmes created are additional and do not take the place of national promotion. Furthermore, the EEC co-operates with the countries and the regional authorities in these accomplishments. Thus, the growth of a suprabureaucracy is avoided and it is made easier for the German states, for example, to play their constitutional part.

2 Regional Development in the EEC: the Example of the Federal Republic of Germany and the State of Saarland

The importance attached to the principle of subsidary which the president of the European Community Delors has successively urged upon the EEC Commission and also on member countries of the EEC enables the governmental units and the states of the Federal Republic of Germany to implement regional innovation, and therefore economic promotion, according to the power of ideas and finances. Thus, for regions like the Saarland the EEC support is not to be underestimated.

2.1 The Foundation of European Regional Policy

Article 2 of the EEC Treaty instructs the European Community to achieve a harmonious development within the European Community as well as a steady and balanced expansion of the economy through the step-by-step approach of the economic policy of the EEC members. In accordance with the preamble, the distance between the well-endowed areas and those with fewer amenities is to be decreased. Through expansion of the European Community with the addition of Greece and then Spain and Portugal, the imbalances in the economic strength of the single regions of the European Community have increased, the North-South incline has become apparent. Therefore, the members of the EEC have committed themselves, as laid down in the Single European Act, to essentially increase the financial means for structural development in the single fields and to direct 80% of it to the strongly disadvantaged southern regions and to Ireland. This goal is being administratively realized through the reform of the structural funds (regional and social funds and the agricultural fund).

Furthermore, it should not be overlooked that problem areas can also originate in relatively rich areas - e.g. because of sectoral developments - that make a restructuring and adjustment and therefore special promotion necessary. This goes, for example, for the coal, iron and steel, textile or footwear industries and harbour regions, where the shipbuilding industry has nearly completely died off. Therefore, the regional funds of the EEC have a continuing significance, e.g. in the Federal Republic of Germany, for the improvement of the infrastructure in order to support the settling of new businesses and the expansion of existing businesses in such regions. Therefore, not only the restauration of contaminated sites but also the transfer of technology are promoted by the EEC through, for example, the steel-location and the RESIDER programme.

2.2 Technology Transfer and the Promotion of Innovation

The transfer of technology is of immediate significance for the restructuring of neighbouring regions. Transfer of technology often refers to making results from research and development in universities/colleges and other research facilities known, so the results can be applied to production faster, partially with the help of the EEC, but also most of all with the help of the federal government.

In this context I want to mention some examples for innovation in the state of Saarland. At the computer science department of the University of the Saarland a group of outstanding scientists work. A fifth faculty including computer sciences, electronics and material science (most of all: new materials) is to be established in accordance with a recent amendment of the university laws. The achievements of the computer science departments find general recognition, so therefore the Max-Planck-Society has decided to open a first Max-Planck-Institute for computer science in Saarbrücken. A centre for artificial intelligence is located at the universities of Saarbrücken and Kaiserslautern.

The Saarland has traditionally maintained tight contacts with the neighbouring Lorraine. The German-French university institute (DFHI) was founded with lectures being held in

Saarbrücken as well as in Metz with the goal that the students receive a German and a French graduation certificate when they finish. Next to the department of management sciences, a joint department of management and engeneering, mechanical engineering and electronics, and computer science has now been introduced to this German-French training programme. I may add that the graduates, without exception, immediately find excellent jobs.

2.3 Europe-Led Regional Policy

Let me make an important remark regarding the general political framework of the EEC regional policy. The demand for balanced regional development also includes that means for promotion from national sources are to be granted only within a framework determined by European Community laws. New distortions, caused by national measures of support should be avoided. Affluent regions, which naturally have more money at their disposal for new locations, are not supposed to use this beyond the scopes determined, to deliberately disadvantage poorer regions and to exclude them from a harmonious development. On the other hand, it must also be avoided that poorer regions cause dumping situations with inadmissable measures of promotion that would work to the disadvantage of other parts of the Community and to the businesses that are active there. In this sense the EEC plays the part of a referee; it has to provide the conditions and has to care for their observance.

Regionalization in no way hinders European integration. Rather regionalization makes possible a "unity in variety" and thus precludes the notion that the European states have to renounce their cultural inheritance in their struggle for the realization of political and economic unity. Present examples in Eastern Europe show that patronization in cultural matters by a central authority far away can indeed contain a considerable explosive political force. As the example of the Federal Republic of Germany shows, the other way around, the decentralization of decision-making, leads to competition among the regions also for economic development - a competition that has contributed a lot to the prosperity of the Federal Republic of Germany.

3 Experiences with Frontier-Crossing Co-operation

The regionalization of border regions is of special significance, because a border-crossing co-operation beyond the single-state borders is made possible. Under this aspect I would like to point out some principles of the border-crossing region Saar-Lor-Lux. It is about a border-crossing co-operation between the state of Saarland, the region of Lorraine and the country of Luxembourg, which corresponds to a practical need in the border regions and which also includes parts of Rheinland-Pfalz and Belgian-Luxembourg.

The border-crossing co-operation in the region Saar-Lor-Lux has been accomplished on several levels. There is an interregional parlamentarian board (IPR), where the presidency and the parties of the participating parlamentary assemblies are represented. The business management of the IPR was granted to an Inner-Community Regional Institute (IRI), based in Luxembourg. The ministry board of the State of Saarland holds joint sessions with the government of Luxembourg and the presidency of the region of Lorraine. A union of the municipalities is founded as ComRegio. The Chamber of Industry and Commerce and the Chamber of Handicrafts of the state of Saarland, Lorraine and Luxembourg have regular joint sessions. An interregional trade union board meets several times a year in the region.

The Saarland and Lorraine have a comparable economic structure. The structural crisis of coal and steel leads to considerable unemployment. There is an interest in the increasing extension of new industrial branches in the field of innovative technology. Regarding computer science, which is also well represented in Metz, a connection of the research facilities through a border-crossing glass-fibre-cable has been already decided on. The Saarland has participated in the formulation of the next five-year-plan for the region of Lorraine. The ministers of the government of the Saarland are having talks on specialized subjects with the competent authorities of the neighbouring French region. A multi-language brochure has been produced for the presentation of the Saar-Lor-Lux-region. Many other initiatives have already been initiated. But it is a process of development; there is still a lot to do in the future.

1.2 The Role of the Commission of the European Communities in Regional Development Policies in Europe

T. Joseph Mulcahy

The evolution of the European Community's efforts to correct its economic disparities has not achieved the success originally anticipated. The institution of the Regional Fund in 1975 has not been sufficient in itself to carry out the daunting task of overcoming the root causes of regional imbalances.

1 The Single European Act

The enlargement of the Community accentuated these problems and coupled with a complexity of economic, social and political pressures led to the need to revise the EEC Treaty by means of the Single European Act. Primary features of this revision are the completion of the Internal Market by the end of 1992 and explicit recognition of the need to redress regional imbalances and regenerate declining industrial areas in a more fundamental way. This reformed approach was effected by amending the character and operating rules of the existing Structural Funds (European Agricultural Guidance and Guarantee Fund, Guidance Section, European Social Fund, European Regional Development Fund). The Regional Fund is specifically identified as the major instrument among the three funds which will act in future in a far more concerted manner to achieve economic and social cohesion as set out in the "Framework" Regulation approved by the Council of Ministers on 24 June 1988.

The European Council in February 1988 had already decided that given the advent of the Single Market a redoubling of resources, i.e. from 7,000 mio ECU to 14,000 mio ECU, was required to overcome the regional disparities in productivity, employment and income identified by the Commission in its Third Periodic Report.

2 A Total Systems Approach to Regional Development

The permanent strengthening of local economic performances implies a total system approach to the development of the regions and the planned exploitation of their resources, human and physical, public and private, as well as involving those aspects of other Community actions with regional connotations e.g. SME Task Force programme, EUROPARTENARIAT, SPRINT, COMETT, ESPRIT, BRITE.

Implicit in this approach is the recognition that the cofinancing of infrastructure investment must be accompanied by concentrated attention on the less tangible and more difficult task of establishing regional development strategies and allied ranges of services. These would be oriented towards identifying and analysing the strengths and weaknesses of the region, such as native entrepreneurial capability, quality and skills of the labour force, number and type of existing SMEs, their likely capacity to innovate, management abilities, training facilities, sources of technical expertise, nature and extent of business development services, market research, financial facilities, national and international network opportunities.

The thrust of this exercise is to establish a global strategic plan for the development of the area in which all the principal local interests would be involved, which would build upon the strengths and rectify the weaknesses and through networking link with and learn from the experience and knowledge of other areas throughout the Community.

3 Local Global Strategic Plans for the Support of SMEs

The Commission, in its Guidelines for Community Regional Policy indicates its intention to support the establishment of local global strategic plans aimed at creating and developing small and medium businesses in the industrial and artisanal sectors drawing upon public and private initiatives.

3.1 Services for prospective entrepreneurs

While more widely available and better quality services for prospective entrepreneurs and existing businesses will be promoted in general, priority financing will be afforded to those concentrating on:

- The promotion and detection of entrepreneurship
- Business creation based on management skill development and business plan preparation

- Innovative techniques in management and technology
- Marketing and export possibilities
- Corporate financing
- National and international networking

The aim is to establish in the less developed areas a range of quality business development services easily accessible and as complete as possible.

The improvements thus brought about in the character and quality of start-up businesses and existing firms seeking to adapt to technical and market changes will facilitate their access to financial resources. Indeed, simply making available information on the existence and methods of drawing upon these resources is of considerable help.

3.2 Strengthening of SMEs

There is, nevertheless, a need to strengthen the financing possibilities for SMEs in regions to which the Commission plans to respond, by:

- Establishing or participating in loan guarantee schemes, venture and seed capital funds
- Cofinancing interest subsidy schemes
- Contributing to the operating cost of such schemes
- Introducing alternative financing arrangements to traditional methods e.g. leasing

In certain specified instances where no aid scheme exists, direct aid to SME investment may be made available, e.g. to very small businesses in rural or touristic areas.

Finally, small-scale infrastructure and equipment essential to the proper functioning of the aforementioned services may be cofinanced.

4 Regional Development Organisations

In the context of achieving the partnership between regional, national and Community authorities which is an inherent part of the reform of the structural funds and to ensure the effective implementation of the global strategy embracing these aforementioned services and financial aids, the Commission plans to encourage, in agreement with the member state, the constitution of regional development organizations capable of undertaking the organization and management of all the functions involved viz:

- Regional strength and weakness analyses
- Promotion of the region, its enterprises, products, exports

- Stimulation and exploitation of local entrepreneurship
- Comprehensive services for the creation of new businesses and the expansion of existing firms
- Corporate financing facilities

Such an organization may be public, public/private or private non profit-making. The Commission may entrust to regional development organizations conforming to the above criteria, the management of global grants which would be used to carry out the aformentioned functions. A contract would be concluded between the Commission and the organization concerned specifying the conditions of the grant and the work programme.

Cooperative networking must be set up so that knowledge and experience with the global strategies being planned, and indeed other initiatives, are exchanged and an overall coherence in regional planning approaches ensured. In the Community context, joint operations between regional policy measures and other actions being promoted on a Community-wide basis are being reinforced with a view to generating feedback mechanisms between the prosperous and less prosperous areas which can be mutually beneficial e.g. Europartenariat jointly managed and funded by Regional Policy and the SME Task Force which seeks to set up joint cooperation agreements between selected firms in the less developed areas and their more prosperous neighbours.

5 Conclusion

In conclusion, at the risk of repeating what has been said before in many quarters, the greatest single resource the regions have is its people and their abilities. The measures being launched under the new structural funds regime are aimed at making sure that the capabilities of the people of the regions and their organizations receive the maximum support and encouragement from the Community, which represents their fellow citizens, for their endeavours to overcome their disadvantages, develop strong local economies and achieve a better way of life.

1.3 A Businessman's Perspective of Regional Policy

Gwyn Jones

My paper looks at regional policy but more importantly at the application of part of regional policy as viewed by a businessman. I emphasise businessman as that is what I am. Even though I am now the Chairman of the Welsh Development Agency (WDA), my roots are in business and at the end of my three year term I shall return to business. My background is the standard route for many an entrepreneur. I was an academic, a computer scientist, and a middle manager. Then I started my own business. Five years later I sold it. I retired and travelled the world, painted watercolours, read books, ate my fingernails, got thoroughly bored and went back to business. Some months ago the Secretary of State for Wales asked me to become chairman of the WDA: I accepted enthusiastically. You can see I am a poacher turned gamekeeper (for a while) or in the words of the 1960s song "I look at life from both sides now". I must admit that to the businessman the phrase regional policy sounds dreary. It smacks of academic debate and grand master plans. I'm glad to say I come from a generation that intensely dislikes grand master plans. If regional policy is to be meaningful to me, and many like me, it must at most be a framework and it must encourage individual enterprise and innovation to flourish. I have come to realise that to fully exploit regional policy a region must ensure that those people tasked with delivering the policy must themselves be enterprising and innovative. All too often this is not the case.

Let me show you what I mean by reflecting on the evolution of the WDA and I'll start by reviewing the problems that have had to be tackled.

1 The Economic Situation of Wales

Wales is one of the four countries in the United Kingdom. It has a population of some three million and its capital, Cardiff, lies some 150 miles west of London. Traditionally the economy centred around three industries: coal, iron and steel, and the shipping associated with these. In the 1920s half of the entire working population (50%) was employed in these industries, reflecting not only their dominance but a strong employee culture. By the late 1970s this figure of 50% had been reduced to 4% as these industries were decimated, e.g. the number of coal pits has fallen from over 200 to 9.

Needless to say in the 1970s, when the final tidal waves of demanning hit the economy, unemployment soared to 17% and the full ramifications of having an employee culture dependent on three basic industries became all too evident. But it is now 1988 not 1978 and the miseries of the past decade are behind us. We look with real optimism to the next decade. Unemployment has fallen for the last 24 months and continues to fall: figures for self-employment are well above the UK norms reflecting a switch from the employee culture to the enterprise culture. Company formations are at an all time high: company survival is the second highest of all the UK's regions. Inward investment continued at record levels, a financial services sector is being created with the most prestigious names participating. I can go on and on. Of course, there is much more to be done but so much has been achieved already and confidence abounds. But the route has not been an easy one, it has been marked by "glorious victories to great defeats". There were false dawns but there was also progress and the right solutions eventually evolved.

2 The Welsh Development Agency before 1980

Originally the only vehicle available in the UK to tackle problems of low prosperity was the regional policy which originated in the 1930s and was based on giving government grants to companies relocating into areas defined as being economically depressed. Although this policy was successful at a general level, it was felt that the particularly severe effects of economic restructuring in Wales and Scotland merited the creation of regional development agencies to supplement the traditional approach. In 1976, therefore, the Welsh and Scottish development agencies were established. The purpose of the Welsh Development Agency is to help regenerate the economy of Wales. The 1975 act outlined the three main functions of the WDA:

1. To further the economic development of Wales and in that connection to provide, maintain, or safeguard employment
2. To promote industrial efficiency and international competitiveness
3. To further the improvement of the environment in Wales

The WDA operates alongside the government department; the welsh office, which administrates regional grants.

So here was a new organization created to tackle a declining economy. What did it do? Initially the agency focussed on building advance factories and reclaiming areas of industrial dereliction. The hope was that government grants would entice companies to move from more prosperous areas to fill these factories. In fact, the approach became build, build, build, but as few were taken up, the countryside became littered with empty factories. This did very little to enhance local confidence. Grants were given to ailing businesses or ones that turned out to be "fly by night" in a desperate attempt to fill these factories. This helped no one in the long term - business failed leaving creditors unpaid.

3 The Welsh Development Agency after 1980

In 1980 a new approach was adopted. This was based on the premise that if we cannot attract enough industry into Wales then we must encourage the businesses in Wales to grow. The trouble was that the people in the agency did not really know how to do that because they had no experience of business. They turned to what they felt was the safe route and focussed entirely on those businesses already in existence. Often they tended to be remnants of the past. Not ready to change to products for the future and so this policy of offering them bigger and better factories was not successful. Ironically the real solution was emerging at this time, as a number of factors had interacted to spawn the beginning of a new enterprise culture - of which I was part. But the policy implementors could not yet relate to this enterprise culture. It was alien to Wales and alien to them. They did not believe that the future lay with this diverse range of emerging businesses. They believed that the existing businesses held the key. So the new corporate advice resources and attention were concentrated on those and the new wave were left to largely fend for themselves with the minimum of help. But at least a business advice structure had been established - the business development unit or BDU as it became known. The trouble was it was manned largely by untrained people. They came from the public sector yet held themselves out to be advisers, mentors to, even gurus to small businesses. Frankly many of them were unable to do the job and hence created a bad local reputation for the agency. Business people like me felt the whole policy was a waste of time and money - with hindsight we were wrong it was the way it was being implemented that was wrong.

Time moved on and the then Secretary of State recognised that a change of direction was necessary, and so he recruited two businessmen to run the agency, two people who were active in business. This led to the transformation of the agency spearheaded by David Waterstone then, and still now, its excellent chief executive. He recognised the obvious, that if the agency was to have the confidence of business people then it had to speak their language and understand their problems. He also recognised that the new wave of entrepreneurs were still in business and were still growing in size and number. So this became a target market - what could be done to help these grow and to encourage more to start.

A new strategy was evolved whereby the agency offered services in three complementary areas, business advice, investment and property. A metamorphosis was started, commercial people were recruited and others were retrained. The agency had to be a player in each of the three services: it had to offer advisers, investment money and premises to let. By being a player it would stimulate the involvement of the private sector through co-operation and sometimes competition. Let me emphasise this point by looking at investments. Initially the agency just gave advice about sources of equity and debt finance and the private sector responded slowly. When the agency set up venture funds of its own the private sector responded much more positively.

4 The New Approach of the Welsh Development Agency

This new approach has been highly successful. It started slowly but gathered momentum quickly, for example:

- Wales now has 15% of the workforce in self-employment
- 52,700 company formations were registered since 1982
- The business advice division has record activity
- The agency has investments in over 500 companies mostly in the equity GAP
- The agency has 18,000,000 square feet (1.8 mio. square meters) of premises of which 94% is occupied and the forward build programme is committed
- Rentals have reached private sector levels in many areas

But, of course, even this new strategy has had its successes and failures and we must constantly review our tactics. It is relatively straightforward to set up an organization to give "start-up" advice - advice on starting your own business. This has been and continues to be successful but challenges emerge.

1. As the spirit of enterprise accelerated, more and more public bodies wanted to get involved in giving business advice. Sadly these services are not always complementary and sometimes are in competition. This is obviously not good for the entrepreneur. We are tackling that.

2. As businesses begin to grow they face new challenges and need new advice - but this advice is now sector dependent. One organization cannot offer expert advice (continuing advice not a single interview session) on everything in a diverse economy.

If you try you do no one any favours, you have to choose your sectors of strength and concentrate on those. This we have done not only with advice but with sector initiatives in film, food, furniture, craft and several others.

3. There is an obvious tendency towards high technology and this is a tendency that we want to encourage. This was so strong that we set up a special organization - Wintech. Wintech was created in 1984, to develop the technological base of existing Welsh companies whose future survival was seen to depend on their ability to exploit technological innovation. Wintech began by providing practical help to companies with problems requiring a technological solution and by providing information on support facilities for industry at universities and commercial research and development organizations. The WDA's company investment strategy in recent years has given priority to technology oriented firms. Over a third of WDA venture capital goes to technology sectors. But note it's not all high technology and we must not forget the low technology industries.

4. Challenges also evolve in our investment division where the bulk of the investments are in the equity gap and represent risk capital. It is a major challenge to find people who can make an economic return on funds invested and yet accept public sector pay rates.

5. Challenges exist in the property department as we have to be ever more innovative in devising schemes for increasing the private sector contribution to our schemes.

The challenges of advising growing SME's continue.

1. Their ever-growing thirst for investment capital. We have addressed this by attracting a financial services sector into Cardiff.

2. There is little research available to help an SME measure its own success. In my own business we found it easy to compare our own performance with our own performance last year, but how well were we doing compared with other companies in similar sectors. We could not solve that question despite enlisting the support of a business school.

3. Culture gap may be emerging whereby SME's find it increasingly difficult to attract quality personnel from large companies. It seems that the view is emerging in some big business that if you have been active in small business you may not fit into the culture of big business.

4. Entrepreneurs sell out often to public companies and the control of the entrepreneur's company moves out of the region.

I quote these as examples to show that you must not have retired bank managers, retired accountants etc. as your key business advice staff. The challenges facing SME's are dynamic and need people commited to finding solutions not just those who merely want to draw on their experience. In Wales we enlist the support of the new breed of entrepreneurs who are pleased to work with us. These are the new breed: the relevant not just the eminent.

5 Conclusion

In this paper I have reviewed the evolutionary path we followed to stimulate the indigenous business base. I have not discussed our inward investment arm Winvest, which now gets over 20% of all inward investment projects coming into the UK, while we have only 5% of the population. It's latest success was the £750M Ford plant. I have not discussed our policy of subregional targeting; far too many people think of a region as one homogeneous area, while often it is several diverse subregions. I also have not discussed our approach to 1992.

In conclusion let me say that regional policy can be justified on economic, environmental and social grounds. The approach in Wales of combining targeted discretionary government grants together with the services of the WDA and ensuring that those services are relevant to business needs and delivered in business language is ensuring that growth companies rarely fail to secure the help they need. I am confident that my own sad experience back in 1982 could not happen in 1988. For my staff it has been a long hard road but now they are blessed with the early fruits of success. The economy has been reshaped, it is vibrant, the enterprise culture flourishes. There is more to be done, much more, but as always it is easier when you can see the rewards of what has been done so far.

2 Regional Disparities in Europe

2.1 Structural and Regional Differences in Economic Development in the Countries of the European Community

Joachim Genosko

1 Introductory Remarks

The assignment of regional policy competences to the European Commission in Brussels and the coordination task associated with this put the question of European wide regional development differences in the centre of our considerations. As already pointed out in a publication by the Wissenschaftszentrum Berlin differences in regional development have something to do with the dependence of regions on industries which are heavily affected by structural change. Such industries include agriculture as well as coal, iron and steel and the textile industry. Thus, it seems to be very difficult to separate structural and regional development differences from each other[1].

2 Main Indicators

I will examine the development of regional differences using major indicators, which not only contain typically economic indicators, but also indicators describing the respective regional living conditions in a broader sense. In general, a region's economic well-being is asessed in the first place on the basis of relative regional income. Income differences can be measured with the aid of various income indicators. I use the Gross National Product (GNP) per employed person because the GNP or the regional income are determined on the basis of resident status, irrespective of place of performance. Thus, we are more likely to get an indicator for the regional level of prosperity. Also the reference to the employed persons makes differences in the population structure (especially in the birth rates) less visible. In this context several questions are of interest: What do the income differences at the EEC-level look like? Which income differences arise within countries? Have the income differences enlarged or decreased over time?

Table 1: Coefficients of variation and mean values of income calculated on the basis of index figures (EUR 12 = 100) [a]

	Coefficients of variation	Mean values of income
Over all regions of the 12 EEC member countries	0.34	93
Over all regions of the 6 EEC founding countries	0.18	114

a Basis of calculation: ECU 1983

Source: Eurostat 1987; own calculations

From Table 1 it can easily be seen that the relative dispersion (i.e. the coefficients of variations[2] among the 12 member countries are clearly higher (nearly twofold) than among the six founding members of the EEC. In connection with the mean value of income we can draw the conclusion that the newly joining members of the EEC deviate in the direction of a below average income. If we restrict the calculation of the mean income values to only EEC countries with below average income values, then the mean value amounts to 62. Therefore, we can draw our first conclusion, that among the regions of the EEC considerable income differences exist, where the regions with below average GNP per employed person concentrate (with the exception of Ireland) on the Mediterranean countries Greece, Italy, Portugal and Spain. Portugal and Greece show by far the biggest income deviations towards the bottom. Within the latter countries the income differences are relatively low; in Greece e.g. the corresponding coefficient of variation has only the value of 0.20.

Table 2: Coefficients of variation and mean values of income calculated on the basis of index figures (EUR 10 = 100) [a]

	Coefficients of variation	Mean values of income
Over all regions of the 10 EEC member countries	0.34	97
Over all regions of the 6 EEC-founding countries	0.20	111

a Basis of calculation: ECU 1980

Source: Eurostat 1984; own calculations

A comparison between Table 1 and Table 2 shows that the relative dispersion over the entire EEC does not change because of Spain's and Portugal's accession to the EEC; the relative disperson over the 6 founding countries, however, has decreased by 10 percentage points. But the extension of the EEC has led to a clear spreading of the income mean values; the corresponding difference increases from 14 index points to 21 index points, first and foremost because of the drop of the income mean value for the entire EEC. Apart from income differences clear distinctions in the labour-market situation arise. While in the 4th quarter in 1986 the unemployment rate in the whole EEC amounted to 10.9 percent on average, for the countries with below average income it was 13.3 percent. However, this latter rate was first of all a consequence of the high unemployment in Ireland and Spain. But in this context we should not fail to notice, that also within EEC-countries with relatively low unemployment there are regions with rates clear above the EEC-average: for instance the region Bremen in the Federal Republic, the region Mediterranée in France or the region Groningen in the Netherlands.

We have run a simple regression equation. The regional unemployment rates represents the dependent variable. The RHS variable is formed by the sum of the regional percentage shares of employees from sectors like coal and iron mining, processing of rock, stone and related mineral products, manufacturing and processing of glass, vehicles construction (without automotive construction), and the textile, leather, footwear, and garment industries. We then obtain the following estimates (t-values in brackets):

$$\mathbf{URATE} = \underset{(3.513)}{7.393} + \underset{(3.163)}{22.152}\,\mathbf{PS} + \mathbf{u} \qquad (1)$$

$$\mathbf{R^2} = 0{,}385$$

As we can see from Equation (1), the RHS variable is highly significant. Equation (1) represents a cross-sectional estimation, which contains the respective three regions with the highest unemployment rates in the 4th quarter in 1986 of the different EEC-countries. Since the above mentioned industries are more or less crisis-ridden sectors, we can conclude: unemployment is among other things affected by an existing economic structur susceptible to crisis.

The interregional migration flows provide another indication for structural and regional faults: for instance, the northern and middle regions in the FRG, the Spanish regions

Cataluna and Pais Vasco, the French regions Ile-de-France, Nord-Pas-de-Calais or Lorraine, the region Antwerpen in Belgium or the region Noord-Holland in The Netherlands show considerable negative balances of migration. These regions are also characterized by crisis-ridden industries. OLS cross-section estimations contain immigration into regions with the highest positive balances of migration per country as the dependent variable in each equation and as RHS variables either the labour market situations (expressed by unemployment) or the economic structures or indicators for the existing living conditions (living conditions measured by health personnel and equipment, number of full-time pupils in different types of schools, quantity and quality of dwellings, household equipment with cars, telephone and TV as well as household electricity consumption). We have run three simple regressions, which come to the following estimates (t-values in brackets):

$$\mathbf{IM_r} = \underset{(1.739)}{24477.14} - \underset{(1.430)}{1281.25}\ \mathbf{URATE_r} + 19498.46\ \mathbf{RATIO} + \mathbf{u} \qquad (2)$$

$$\mathbf{R^2} = 0.045$$

$$\mathbf{IM_r} = \underset{(0.122)}{-1173.93} + \underset{(3.503)}{203639.43}\ \mathbf{ES_r} + \mathbf{u} \qquad (3)$$

$$\mathbf{R^2} = 0.305$$

$$\mathbf{IM_r} = \underset{(0.087)}{783.11} + \underset{(3.524)}{918.55}\ \mathbf{LC_r} + \mathbf{u} \qquad (4)$$

$$\mathbf{R^2} = 0.307$$

Let **IM** be the immigration flows and **URATE** the unemployment rates. **RATIO** corresponds to $\mathbf{URATE_r/URATE_n}$. **ES** describes the different economic structures and **LC** stands for the living conditions. **r**, **n** and **u** denote the region, the nation and the error-term respectively. Each EEC-country is involved with three regions. From equation (2) it is clear, that immigration ist not influenced by the labour-market situations, even if the coefficient of **URATEr** has thc right sign. This result is consistent with other empirical literature[3]. Simultaneously the impact of **LC** on **IM** seems to be somewhat stronger than the impact of **ES** on **IM**. This lets us suppose, that living conditions also play an important role with respect to regional development and not only economic factors[4] - an outcome, which is meaningful in designing a successful regional policy.

3 Structural Differences as Determinants of Regional Development Differences

Since I have several times pointed to the importance of the economic structures for regional development differences, it seems reasonable, to consider the relationship in more detail. First, it is necessary to distinguish different types of regions following a differentiation developed by the Institut für Weltwirtschaft at the University of Kiel[5]. I distinguish four types of regions: lagging regions, developed or mature regions, highly developed regions and old industrialized regions.

The industrial structure of lagging regions is characterized by labour-intensive industries and labour-intensive production processes in industries, which are more capital intensive. Industries, which play a role in this context, are e.g. food processing, processing of rock, stone and related minerals, and the timber, textile, footwear, leather and garment industries. Indeed, these industries measured by their shares of employed persons show in Ireland as well as in the southern regions of Italy and France above average percentages compared with the values of FRG or The Netherlands, as we can infer from Table 3.

Table 3: Share of employees in the food processing, processing of rock, stone, and related minerals, and in the timber, textile, footwear, leather, and garment industries as share of overall employment in manufacturing industry (1984) [a]

Ireland	0.330
Southern Italy	0.321
Southern France (Sud Ouest, Mediterranee)	0.269
Federal Republic of Germany	0.214
The Netherlands	0.201

a There were no corresponding data available for the regions in Greece, Spain and Portugal

Source: Eurostat 1987; own calculations

Developed or mature regions are characterized by a certain industrial diversification. In this kind of regions industries like the metal-working, the mechanical engineering or electrical engineering industries play an essential role. In developed regions with agglomeration centres chemical industry is often added to the industrial "picture". In spite of data problems we can empirically underpin our assertion.

Highly developed regions are characterized by a demand for highly skilled workers, i.e. they require a high utilization of human capital. Industries which mark such regions are e.g. electronics industry, the production of hardware, the chemicals fibres industry as well as publishing and printing. In this connection, Table 4 shows which regions in particular have such industries within their borders.

Table 4: Share of the employees in the metal-working industry, in the mechanical engineering industry, in the electrical engineering industry, and in the chemicals industry at the overall employment in the manufacturing industry (1984)

Bayern	0.334
Baden-Württemberg	0.391
Hessen	0.388
Emilia Romagna	0.288
Bassin parisien	0.342
Nord Ouest	0.348
Federal Republic of Germany	0.376
France	0.301
Italy	0.236
Ireland	0.151

Source: Eurostat 1987; own calculations

From Table 5 we can see, that highly developed regions are centred around agglomerations. This statement will still be important, if we discuss in more detail one of the causes responsible for the spatial distribution of the economic structures within the EEC.

The last type of region, which we have to consider, are the aging regions. Aging regions, in particular old industrial areas, are characterized by raw material intensive industries, since such industries are naturally located where there are raw material deposits. It is true, that such raw material deposits have been undermined as location factors because of the new interregional and international labour division and because carrying charges tend to decrease over time; nevertheless such industries have kept their weight in the aging regions, even if recently old industrial areas like the Ruhr area or Wales have tried to diversify their economic structures by the aid of environmental technologies. Coal and iron ore mining as well as the industries based on mining like the iron and steel sector dominate the picture of these regions (especially the industries mentioned in Table 5 are underrepresented); see Table 6.

Table 5: Share of employees in the chemical fibres industry, office equipment and computer manufacturers, in precision engineering and optics, and publishing and printing as share of the overall employment in manufacturing industry (1984) [a]

Hamburg	0.185
Ile-de-France	0.190
South East [a]	0.190
Federal Republic of Germany	0.136
France	0.147
United Kingdom [a]	0.133
Ireland	0.132
Italy	0.079

a without producing of office equipment and computers (data are not available)

Source: Eurostat 1987; own calculations

All regions mentioned in Table 6 show an above average unemployment measured on the rate of the respective country. While the countries mentioned in Table 6 had an unweighted average rate of 9.7 percent in the 4 th quarter in 1986, the corresponding rate of the mentioned regions was 12.5 percent, i.e. the regions' rate was about 30 percent higher than the country's rate.

Table 6: Employees in coal and iron ore mining, coaking coal producing, and the production and working of metals as share of overall employment in the manufacturing industry (1984)

Nordrhein-Westfalen	0.137
Saarland	0.258
Nord-Pas-de-Calais	0.144
Lorraine	0.268
North	0.243
Yorkshire and Humberside	0.243
Scotland	0.185
Federal Republic of Germany	0.054
France	0.082
United Kingdom	0.147

Source: Eurostat 1987; own calculations

4 The Relationship Between Structural and Regional Development Differences: Some Remarks

In the literature there are a series of approaches, in order to explain the spatial distribution of enterprises and industries[5]. It would be too long to display all these approaches, nevertheless, I would like to deal with one approach in more detail. This approach stems from the fact, that product and in particular process innovations develop into routine productions over time. In other words, in the initial stage such innovations demand above all a lot of quality manpower, which, however, decreases over time. If we assume that high quality manpower is not distributed evenly in the space, then we can derive from this a certain spatial-economic stuctural pattern. This would mean namely, that for the time being innovative productions arise in the central regions, since there the required workers are available. With increasing routinization of the activities productions spread over the space, till ultimately it arrives in the lagging regions, which tend to be characterized by a high proportion of less-qualified manpower and by a high proportion of unskilled and semi-skilled workers. From a policy point of view a political approach could become apparent insofar, as the quintessence of the conclusion here is: the quality (and in some extent also the quantity) of regional labour markets could be bottlenecks of the regional development[6]. Such a qualitative bottleneck factor can be removed by either a regionally targeted use of education and training measures or by promotion of the relocation of qualified manpower to the peripheral regions. As shown above, however, migration, especially immagration, depends essentially on the regional living conditions; as mentioned such conditions are not always optimal at the EEC-periphery. Nevertheless, the creation of the European single market from 1992 on and the greater freedom of movement associated with this could lead to a new distribution of production within the EEC, to a strengthened shift into the sunbelt of the EEC.

5 A Summing Up

I would like to summarize my central conclusions as follows:

There are great regional differences among the EEC-countries, especially with respect to income possibilities. Within the member-countries greater development differences are primarily found in Italy and in a somewhat weakened form in France. At the periphery of the EEC the intranational development situation is relatively homogeneous at a low level.

Regional development differences are consequences of different economic structural developments. Regions which are underdeveloped, as measured on the EEC-scale, are primarily agricultural oriented and characterized by labour intensiv industries demanding a low manpower quality.

The improved freedom of movement within the EEC could lead to new regional dynamics at least in some regions of the EEC-periphery.

Footnotes

1 See D. Yuill, K. Allen, C. Hull (Hrsg.) (1981), Regionale Wirtschaftsförderung in der EG: Anreize für die gewerbliche Wirtschaft, Königstein/Taunus

2 The coefficient of variation has the standard deviation as numerator and the mean as denominator (see T. Yamane (1966), Statistics - An Introductory Analysis, New York, p. 76)

3 See J. Genosko (1988), Zur konjunkturellen Reagibilität von Arbeitslosigkeit und Beschäftigung, Raumordnung und Raumforschung 46, p. 97-106 and J. Genosko (1988), Regional Labour Market Adjustment Processes, unpublished paper University of Hohenheim

4 See e.g W. Mieth, J. Genosko (1982), Qualitative Polarisierung der Regionen als Folge der räumlichen Selektion der Wanderung und der Arbeitsplätze, in: Qualität von Arbeitsmärkten und regionale Entwicklung, Forschungs- und Sitzungsberichte der Akademie für Raumforschung und Landesplanung, Bd. 143, Hannover and the literature quoted there

5 See C. Krieger, C. S. Thoroe, W. Weskamp (1985), Regionales Wirtschaftswachstum und sektoraler Strukturwandel in der Europäischen Gemeinschaft, Tübingen

6 See F. J. Bade (1987), Regionale Beschäftigungsentwicklung und produktionsorientierte Dienstleistungen, Berlin and Bade's contribution in this volume

2.2 Regional Disparities in Europe: Interpretation and Political Consequences

Jean-Paul de Gaudemar

The concept of regional disparity is one of those useful ideas, particularly in politics, which is ambiguous in theoretical terms. Moreover, this ambiguity is probably simply the reflection of a more basic ambiguity which is inherent in regional economics - or regional "science" a discipline which is identified in the first instance by its supposed scope and not by the specific nature of its methods, and which is therefore based on the premise that such an object, i.e. the "region", actually exists. But this object is hardly scientifically constructed: it is first and foremost the product of administrative practices, which certainly give it legitimacy, but at the same time stamp it with their own particular mark. If regional economics is in search of its own identity it is because by definition, just as much for the sake of efficiency, it is condemned to turn everything to account by borrowing methods which may be usefully transposed: a little spatial economics to take account of the "dimension" of its object; a little development economics because therein lie the real problems at stake; a little industrial economics to refine the analysis, etc. This theoretical eclectcism has some virtues, in particular those related to the analytical power of a quantitative and theoretically better structured geography; but is also has its faults, in particular its poor capacity to predict.

The concept of regional disparity lies at the heart of the epistemological problem which has just been mentioned. Through its many forms it encapsulates the information which may enhance the in-depth knowledge of the regions. But this protean-like nature makes it very arduous, if not impossible, to interpret it. Talk about regional disparities may say both all that needs to be said or nothing at all. It may distinguish between the regions or blur such distinctions. The caution required for any attempt at interpretation, and even more for any attempt at explanation, is thus understandable.

The wealth of literature on the subject provides, nevertheless, for several avenues of thought, which should have great impact on the development of regional policies.

1 Standards and Disparities

By definition, a disparity only has meaning in relation to a standard. In terms of regional disparities, this standard does not always have clear outlines.

It may nearly always be agreed that the principle of the equality of citizens provides that standard. The reduction of disparities then results from a narrowing in the average gaps in terms of activities, standards of living or welfare, etc. Thus, objectives relate to GNP or per capita income, household appliance rates, education levels, doctor/patient ratios, etc. The standard is provided by an average individual: the "regional" average must be as close as possible to the "national" average. In many cases, however, it is difficult to define the standard relatively; for example, as the result of objective constraints, or because of the indispensable nature of a product or service, or an inherently negative situation, or even because "regional" differences reveal irreducible modes of social organization, etc. To quote two simple examples: In mountain areas the installation of radio and TV broadcasting stations is linked less to a standard relating to the number of inhabitants they serve than to the constraints in terms of areas of shadow arising from the physical characteristics of the area. The second example: when speaking of unemployment, it would be absurd to consider that there is a "right" level of unemployment. A policy may consequently aim to bring regional rates into line, not with the average rate, but with the lowest rate, and the lowest as reasonably possible.

The measurement of a disparity requires therefore in the first place a method which shows up the diversity of a phenomenon rather than one which refers back to a standard. This statement applies also to the segmentation of both space and time in analysis.

Let us gloss over the question of time. It suffices only to recall that each diachronic comparison of disparities presumes a permanence in time of the variable used, which is a difficult condition to obtain for long-term series (Aydalot 1985: 157 ff.).

The question of space is perhaps even more delicate. For example, if the spatial reference standard is that of an integrated national market, the regional disparities will be most often analysed as varieties or variations in terms of the national value of an indicator. This approach is obviously suited to the most centralized of states, such as France, and is also used in the most common models of "shift-share" analysis.

In general, these models attempt to identify and define, for a regional indicator, the share of a "structural" component - that which indicates what would have happened if the region had been structurally identical to the nation - and the share of a "regional" component which would point in some way to the particular profile of a "region". The regional component is thus defined as the residual effect which is not sufficiently explained by national development, but in some way as an indirect measure of national disintegration triggered by the existence of the regions; the reference standard for economic integration is therefore more sectoral than territorial. On the other hand, in the comparative analyses most frequently conducted at the international level the reference standard used is most often the nation. Regional disparities therefore appear first as disparities between countries. In this way, INSEE has agreed that the fundamental aspect of regional disparities lies in national disparities, as set out in Table 1 (INSEE[1], 1984:26).

In this way, we encounter two main standards of definition which refer to the most commonly held ideas of the best levels of economic integration: one standard of a mainly

Table 1: Influence of national factors on regional disparities

European regional disparities in:	Fischer's test	National influence
Importance of agriculture	13.2	Considerable
Importance of industry	3.6	Strong
Importance of tertiary sector	6.3	Very strong
GDP (per capita: 1979)	5.5	Very strong
Productivity 1979	49.3	Considerable
Development of employment 1970-1979	4.9	Very strong
Activity rate 1979	45.0	Considerable
Demographic pressure 1970-1990	5.8	Very strong
Migration differences 1970-1979	2.3	Average
Average unemployment 1979-1981-1983	10.2	Considerable

Source: INSEE

1 Institute nationale de la Statistique et des Etudes Economiques (National Institute of Statistics and Economic Studies): the body which publishes most official French statistics and numerous economic studies.

sectoral nature as regards infranational disparities (sector, branch, etc. would therefore be the references in terms of integration); and a standard of a mainly territorial nature (national) as regards international disparities. It may of course be asked what would happen if other standards were considered, in particular, the opposite of the two preceding standards. If we take a nation or a state whose basic organization is of a territorial nature (this is particularly the case in federal states), it would seem legitimate to consider the opposite, i.e. that it is the "region" which provides the reference against which possible residual effects inherent in certain sectors or branches of activity may be measured. Similarly, it might be asked at a time when the money and employment markets are being internationalized what new light might be shed in analyses of international disparities, by taking international activity (for example, at the European Community level) instead of the national level as a reference.

Such ideas hardly seem shocking for a territory such as the United States. Why should they be for the EC, especially with the advent of the single market? However, it is realized that such an upheaval would most probably alter the way in which regional disparities are represented. Despite lacking such results (even though the work carried out by the EC's regular reports on the socio-economic situtation of the regions in the Com-

Figure 1: Reference standards

Reference standards / Regional disparities	Sectoral	Territorial
National	Regional dynamics: residual/structural effect component	Analysis of "federal" effect ?
International	Consideration of internationalisation of sectors and branches ?	Predominance of national effect Synthetic indicator

munity, which make no explicit reference to the national level, would support these ideas), one obvious conclusion may at least be stressed: regional disparities translate the dominant factors of economic integration; these factors function like standards whose historic and political background appear fundamental (see Fig.1).

2 Typologies of Disparities

However, much of the table of regional disparities (Table 1), taken one by one, is often difficult to interpret, and as much as the phenomena observed may be contradictory, different considerations apply when trying to combine or aggregate several indicators. A certain number of structures common to several regions of different countries emerge from such exercises in the form of regional typologies.

The typologies developed so far for the economy alone are already numerous. Over and above the elementary typologies (of the kind centre/periphery, urban/rural, agricultural/industrial/tertiary, etc.) at least three families can be identified.

2.1 Diagnostic Typologies

Here the aim is to identify at a given moment in time the relative situation of an area, region, etc. within another area or vis-à-vis comparable areas. Most often these typologies emerge upstream of political decisions: the choice of regions to benefit from aid depends on the diagnosis. An example is the West German procedure which provides for redefining the zones assisted by regional policy (G.A. zones, Gemeinschaftsaufgaben Zonen, zones whose development is regarded as the common task of federal and state authorities; Allen 1988). This procedure consists of two stages:

In the first stage a "synthetic" indicator is calculated for each of the 179 employment zones by a weighted aggregation of four partial indicators:

- Gross per capita value added (22.5%)
- Gross salaries (22.5%)
- Average annual unemployment rate (45%)
- Composite infrastructure indicator (10%)

In terms of the current map of G.A. zones, which was adopted in June 1986 and implemented in 1987, 74 employment zones out of 179 qualify for federal aid (i.e. 28.8.% of the population) as a result of applying this synthetic indicator.

In a second stage, the procedure identifies the employment zones in which the average unemployment rate over the previous five years is more than 160% of the national average. These employment zones are designated as the G.A. zones. Whether it be a synthetic indicator or a simple indicator based on unemployment rates, the aim of such a procedure is to make the most objective diagnosis possible so as to avoid interminable political debate and also to lay down strict rules governing state financial intervention. In this way several Länder (states) recently lost some assisted zones (Baden-Württemberg no longer has any - proof of its economic success) and others gained some.

As another example, in the regular reports on the socio-economic situation and development of the regions in the Community, the European Commission publishes a classification of regions according to a "synthetic index" which is supposed to reflect the respective severity of their problems. This index, which combines four types of variables (per capita GDP in terms of buying power, GDP per employed person, unemployment rate, and job vacancy forecast; EEC 1987: 21 ff.; Annex 2.2. I-D), reveals quite clearly three classes of region or, alternatively, three Europes:

- Peripheral Europe, including most of southern Europe (Greece, southern Italy, Spain, Portugal) and Ireland (plus a part of Scotland) where the combined severity of problems appears to be the most acute (approximately 30% of the European population)
- Central Europe along an axis London/The Netherlands/southern Germany/northern Italy (plus a part of the greater Paris region) where, on the contrary, the positive aspects of development seem to concentrate (approximately 32% of the population)
- Middle Europe consisting of the rest of the Community, in particular most of France and Great Britain, where the scale of problems appear average, this average, however, can hide major disparities (approximately 38% of the population)

In the same way as the preceding typology allows the Federal Republic of Germany to adjust its financial aid policies for the regions, this typology is intended to play an important role in the decisions regarding ERDF, especially with respect to the reform of the structural funds aimed at allocating the major share of the resources to the first class of region.

These diagnostic typologies are attempts at global analysis of regional disparities, the level of synthetic index set may reflect different situations but their consolidated score is comparable. Their interest is less their analytical relevance than their capacity to provide an argument for political decisions. For this same reason, the problem of analysing territorial disparities undergoes a shift: towards identifying a threshold beyond which political action must be undertaken and towards obtaining political consensus on the method and the level of fixing the threshold, in a word towards a problem of economic policy.

2.2 Structural Typologies

The problem here is to identify the various development problems by characterizing the dominant structures of the regional economy under consideration and the main features of its development.

Numerous examples of such typologies exist. The work was conducted on the occasion of the second regular report into the situation of the European regions (covering the period 1977-1979-1981; EEC 1984: 26 ff.) on the basis of a series of 12 static and dynamic criteria (self-employment, agricultural and industrial employment, GDP, unemployment levels, demographic pressure, GDP variations, employment developments, migration rates, etc.). The European regions could thus be divided into seven categories, so as to identify several broad categories of development problems (Table 2).

At one extreme of the typology, class A contains those regions in which all the handicaps are concentrated, otherwise referred to as structural subdevelopment. At the other end of the scale, class G seems to combine almost all the advantages and includes the central locations in the Community. There is, nevertheless, one negative factor: strongly negative migratory balances which reveal a sharp loss in attractiveness, at least quantitatively. The other classes of regions, which are more composite, reflect again the strong presence of certain national characteristics. For example, class B comprises for the most part those peripheral Italian regions with a weak development potential but a strong demographic attraction; class D groups together those highly industrialized French and British regions where development fails to absorb a strong demographic concentration and where unemployment levels are consequently high; class F comprises basically the German regions (together with Alsace and Luxembourg in a relatively well-circumscribed geographical unit) where industrialization is strong and unemployment is not as high as elsewhere, notably as a result of a less dense demographic concentration. Finally classes C and E appear more composite, if only because the former seems to contain those regions with a high GDP growth rate and the latter includes those regions which are predominantly agricultural but in the process of active industrialization, such as the western and southern regions of France.

Table 2: Regional typology for the Europe of ten (Source: EEC-INSEE)

Class A	Class B	Class C	Class D	Class E	Class F	Class G
Ellas	Piemonte	Saarland	Champagne-Ardenne	Basse-Normandie	Braunschweig	Berlin (West)
Sicilia	Valle d'Aosta	Schleswig-Holstein	Picardie	Pays de la Loire	Hannover	Hamburg
Sardegna	Liguria	Lüneburg	Haute-Normandie	Bretagne	Düsseldorf	Bremen
Campagnia	Lombardia	Weser-Ems	Centre	Poitou – Charentes	Köln	Ile-de-France
Abruzzi	Veneto	Münster	Bourgogne	Aquitaine	Detmold	Noord-Holland
Molise	Friuli-Venezia-Giulia	Kassel	Nord – Pas-de-Calais	Midi – Pyrénées	Arnsberg	Zuid-Holland
Puglia	Emilia-Romana	Koblenz	Lorraine	Limousin	Darmstadt	Brabant
Basilicata	Toscana	Trier	Franche-Comté	Auvergne	Rhein-hessen-Pfalz	South-East
Calabria	Umbria	Niederbayern	Zeeland	Langedoc-Rossilion	Stuttgart	Hovedstads-regionen
Northern-Ireland	Marche	Oberpfalz	Scotland	Provence – Alpes – Coté d'Azur	Karlsruhe	
Ireland		Unterfranken	North	Trentino-Alto-Adige	Freiburg	
		Rhône – Alpes	Yorkshire-Humberside	Lazio	Tübingen	
		Groningen	East-Midlands	Luxembourg (B)	Oberbayern	
		Friesland	West-Midlands	Namur	Oberfranken	
		Drenthe	North-West	East Anglia	Mittelfranken	
		Overijssel	Wales	South-West	Schwaben	
		Gelderland		Oest-For-Storebaelt	Alsace	
		Noord-Brabant		Vest-For-Storebaelt	Luxembourg	
		Limbourg (NL)				
		Utrecht				
		Antwerpen				
		Hainaut				
		Liège				
		Limbourg (B)				
		Oost-Vlaanderen				
		West-Vlaanderen				

Another example is a typology of the French regions based on a sectoral analysis of employed persons from 1968 to 1981 (Salvan, in Uhrich 1987: 39 ff.). The dynamic image provided by this analysis is interesting in that it shows a certain convergence of the regions towards the same development model, based on industrial diversification and the growth of the tertiary sector. Six classes of regions can be identified (see Table 3):

The two typologies posit development "models" which may account for disparities encountered in the analysis of different indicators. Of course, those disparities may disappear in a more global view, but this is exactly what the exercise intends: a disparity in itself cannot mean anything if it is not set in the context of structural development.

Table 3: Six classes of regions

I Predominantly agricultural with the development of intermediate goods
II Agricultural sector still strong but accelerating tertiary growth
III Traditional industries undergoing difficult diversification
IV Predominance of intermediate goods
V Diversified industry
VI Central region, i.e. the Paris region with its intensive tertiary development

I	II	III	IV	V	VI
Basse-Normandie Pays de la Loire Poitou - Charentes Limousin	Langedoc-Rossilion Provence - Alpes - Coté d'Azur Aquitaine Bretagne	Nord Lorraine	Bourgogne Picardie Champagne Centre Auvergne	Rhône - Alpes Alsace Franche-Comté Haute-Normandie	Ile-de-France

2.3 Dynamic Typologies

Finally, certain studies concentrate even more on structural developments to characterize the regions. The approach is less to identify structural states than the structural trajectories of regions. This idea, which is at least implied in the works of Salvan quoted above, influenced, for example, the study conducted by Camagni (Camagni, in Aydalot 1984) on the basis of work carried out on behalf of the European Commission. The principle adopted for this typology is very simple; it consists of comparing regional developments using chronological data on three main critera, employment, production and productivity, estimates for which are compared to overall average growth.

Seven types of situation are thus defined, which are can be represented graphically in relation to the three criteria (see Fig. 2).

It is therefore possible to study regions from an economic development perspective using this typology. Camagni highlights two strong diachronic development models for

Figure 2: Types of regional situation
(Source: CAMAGNI-AYDALOT)

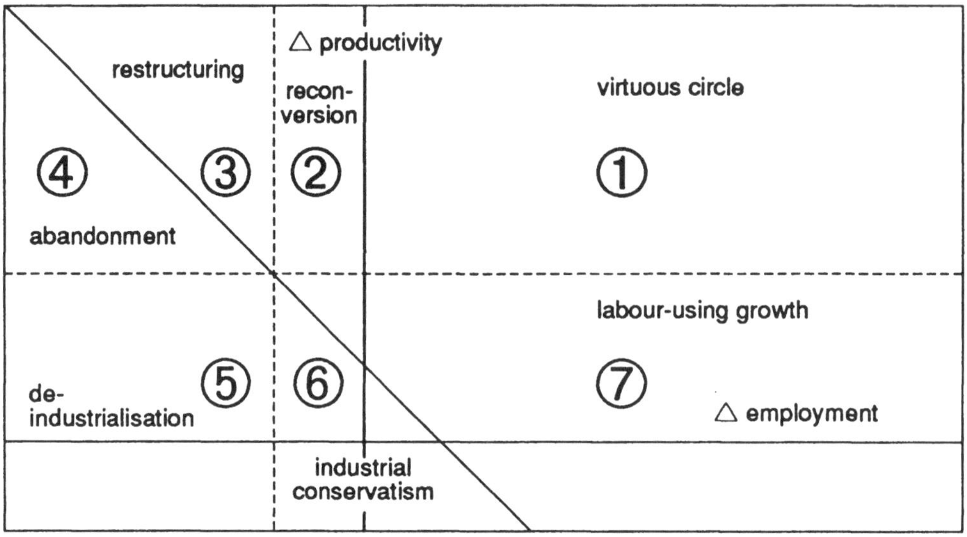

the period 1970-1978: one trajectory, more or less representative of northern Europe, leads from "reconversion" to "deindustrialization"; the other, which could be observed more in the Mediterranean regions, points to recent industrialization leading from "labour-using growth" or from "virtuous circle" to "industrial conservatism".

These two models can be interpreted as territorial types of reaction to a crisis situation. Regional disparities fade in relation to a global behavioral dynamic, which is likely to act in the long-term against these disparities by means of the political choices it triggers.

2.4 Examination of these different typologies leads to several conclusions:

1. Since territorial disparities are less interesting taken separately than together in an overall evaluation of situations, economists concentrate more on elaborating synthetic typologies which are able either to provide a diagnosis or characterize the main features of a strutural development.

2. The analysis of territorial disparities seems largely concerned with identifying the economic policy problems presented by the diversity of territories (international, national, or infranational) and the gaps between them in relation to what might be the ideal "integration" of the territory concerned.

Political authorities seem to be clearly divided on this point between a more or less objective understanding of problems, which is likely to be taken as a fact, or rather as a constraint, on the political debate, and a subjective understanding, where, to the contrary, political factors take precedence over economic indicators. While the Federal Republic of Germany, Denmark and the European Community seem to opt for the first attitude, there is no evidence to suggest the same of other European countries. Therein lies probably one of the most important issues at stake in the coming years as regards territorial disparities.

3. Converging Disparities

Is it possible to identify strong trends via the development of disparities in the way that the preceding analyses present them? Is it possible, in particular, to accept the theory of the convergence of regional economies? Or do new divergences constantly appear? Here, again, is a vital question for economic policy. The theory of convergence thus serves to shore up an argument now widespread throughout the OECD countries. At a time when public spending is being reduced everywhere, one may at least partly justify the cut in regional policy budgets (OECD 1986). The facts are somewhat more complicated at closer sight: convergence is indisputable on many points, and is even quite logical; on others, which perhaps highlight the real weak points of regional development, a sharper divergence emerges.

In the case of certain indicators, convergence cannot be disputed, the trend in per capita GDP, income and salaries is leading in most European countries to a narrowing of the regional hierarchy. In itself, this trend is quite easy to grasp. It reflects the simple dissemination of production and consumption models, which is faster in some countries and slower in others. It is in this respect a fairly faithful reflection of the country's degree of economic and social integration. A high level of economic integration speeds up the dissemination of technical and operational innovations and may thus narrow more quickly the gaps in per capita GDP or salaries. A high level of social integration obviously has an effect on divergences in income and consumption or the level of equipment. Thus in a country like France, the gap in per capita production between regions is much greater than the gap in incomes as a result of the social distribution and transfer mechanisms which characterize French social policy (see Fig. 3).

On the other hand, such trends may be less easily identified in countries with a lower level of integration. The trends for Spain appear less conclusive (Cuadrado 1987): from

Figure 3: Levels of production and per capita income in 1981 (France = 100) (Source: Uhrich)

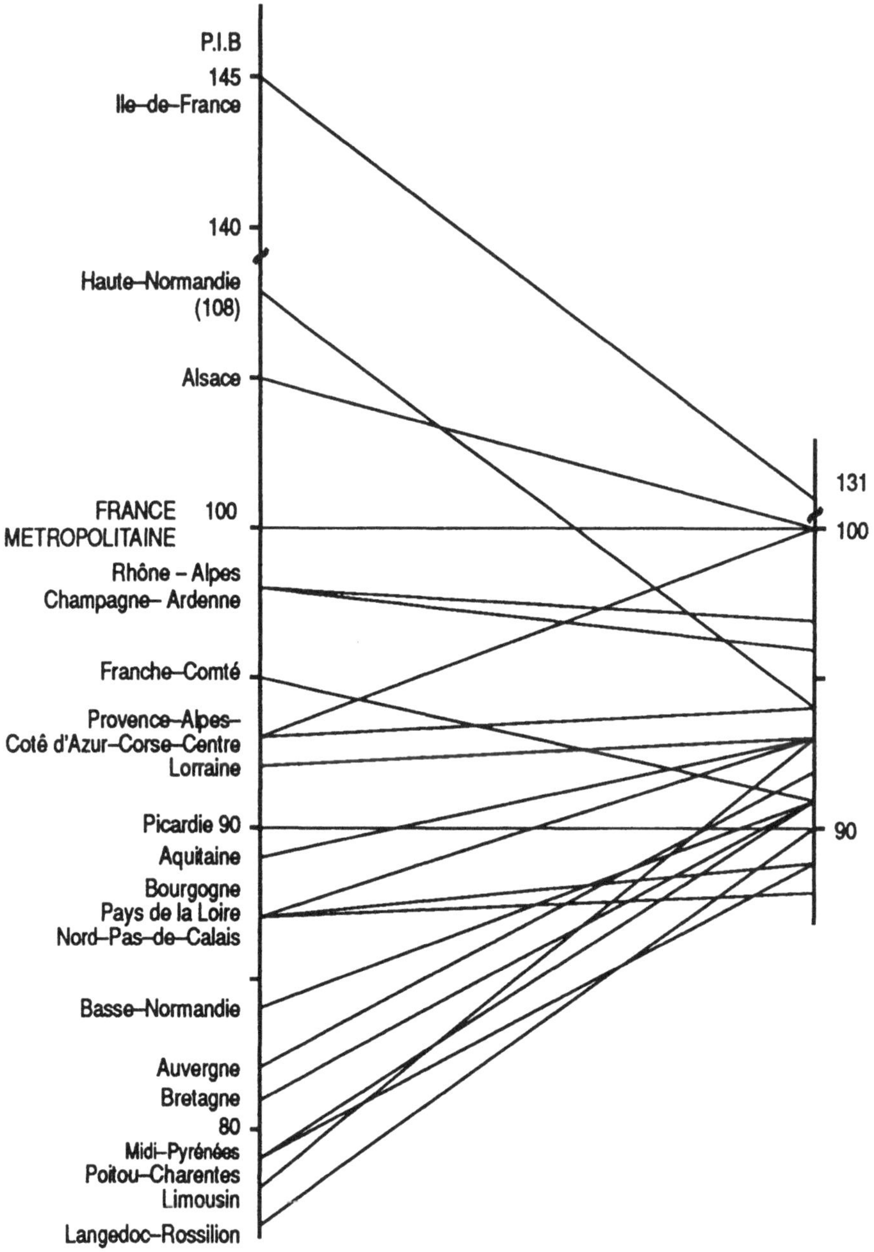

Figure 4: Analysis of regional differences
Source: Cuadrado 1987: 72

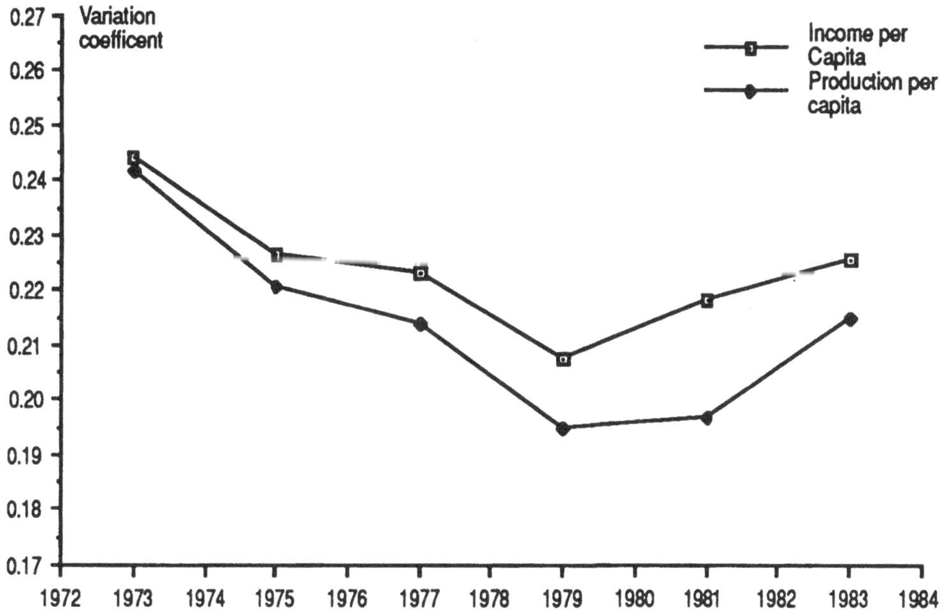

1955 to 1973, per capita production tended to bring about a slight narrowing in the interregional hierarchy and therefore a slight reduction in disparities. However, 1973-1983 reveals a reversal of this trend, both in terms of production and per capita income, starting in 1979; this date, as it were, marks the point at which the economic crisis assumes a more acute territorial dimension (see Fig. 4).

Similarly, when regional disparities are seen from the viewpoint of units, such as the EEC, the trends observed appear necessarily more contradictory as economic and social integration is much slower and even more marginal. It is, however, clear that this is one of the main things at stake in the implementation of the single market. The economic integration advocated must be accompanied at the same time by social and territorial integration which is capable of balancing and harmonizing the expected forms of development.

If it is assumed that the convergence or rather the reduction of territorial disparities is related mainly to the degree of economic and social integration of an economy, the opposite may equally be deduced, i.e. that every difficult economic situation, which reveals the stress points of such integration, is likely to reactivate the relevant disparities. While attacking the weak points of an economy, the crisis emphasises all the possibilities of disintegration. This is basically what has just been noted in the case of Spain, but it may likewise be noted for many other countries, especially as far as the weak link in most European economies is concerned, i.e. the labour market. In the United Kingdom, therefore, the assisted regions suffered relatively fewer industrial job losses than other regions up to 1974 as the national labour market underwent a shift. However, the phenomenon ceases from 1974, and especially from 1978 a clear advantage is created for the wealthy regions in the southeast, thus widening the gap with the other regions (Rhodes, in Martin and Rowthorne 1986, 138-168; see Table 4).

Table 4: Changes in manufacturing employment, by regional groupings, 1966 — 1983 (percentage change in each period) (Source: Rhodes in: Martin/Rawthorne 1986)

	Traditional assisted area regions	Central regions	South-east regions	UK
1966-74	-1.7	-11.2	-8.0	-8.0
1974-78	-10.3	-8.1	-6.0	-7.7
1978-83	-23.2	-27.4	-18.6	-23.8
1966-83	-36.8	-40.7	-29.6	-35.3

A comparable trend may be seen in the case of France. While there tends to be a convergence of many indicators in the various regions, the same cannot be said for the unemployment rate. Unemployment has hit hardest in the most structurally fragile regions since 1975 (see Tables 5a and 5b and Figs. 5-7). This increases steeply the relative imbalance between the regions, to the extent that the divergence seems capable of upsetting the convergences observed in other areas. Unemployment, therefore, appears to act on top of other factors. In France (as in the United Kingdom) dynamic employment policy has tended to come before regional policy.

The weaknesses shown up by the crisis are also quite clear when considering the whole of the European Community; hence the disparities between countries in terms of per capita GDP and GDP per member of the active population. Data in the third regular report show that the trend towards a reduction in these disparities is reversed in 1975 and that since that date the gap has grown slightly but constantly. Less easy to highlight, if only for a lack of available data, the phenomenon seems to be of the same scale as regards the disparities between regions in the Community from the end of the 1970s, as if the economic difficulties had stopped the slow process of achieving territorial balance begun in the Community.

Caution must therefore be exercised when interpreting the facts too quickly or optimistically and concluding that territorial disparities in Europe are narrowing.

Figure 5: Average unemployment rate in 1987
Source: INSEE

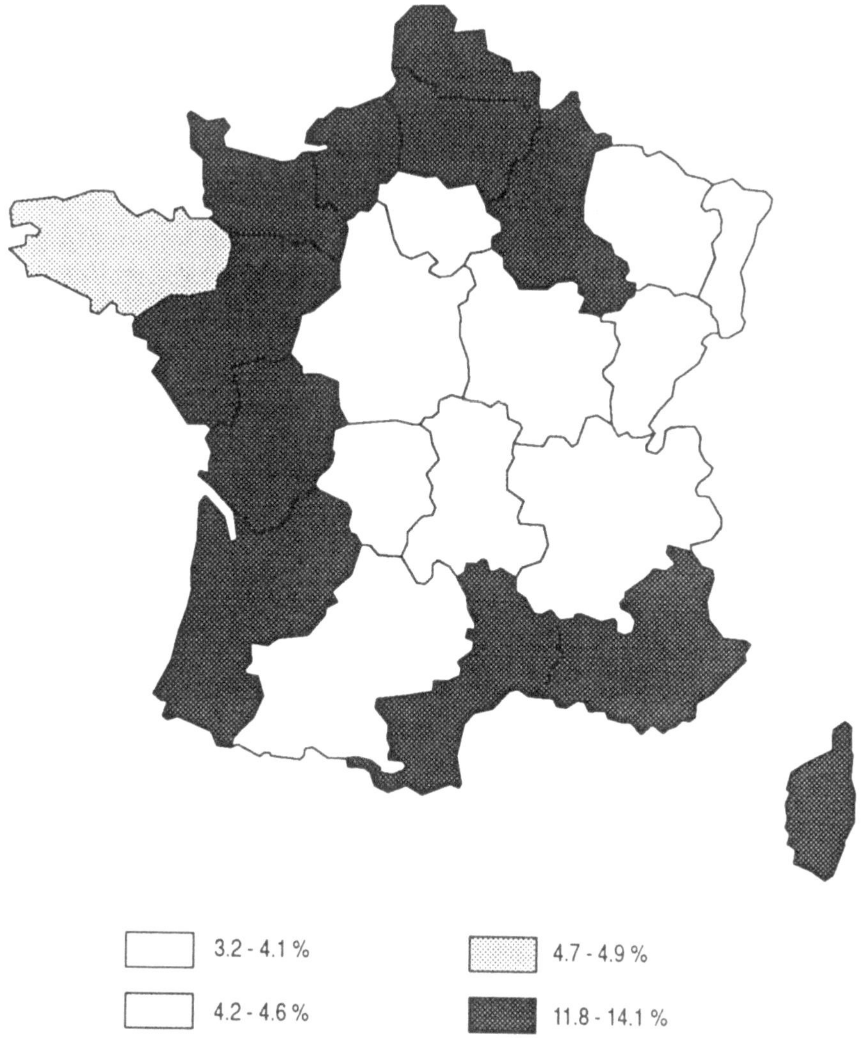

Figure 6: Variation in the average unempolyment rate between 1980 and 1987
Source: INSEE

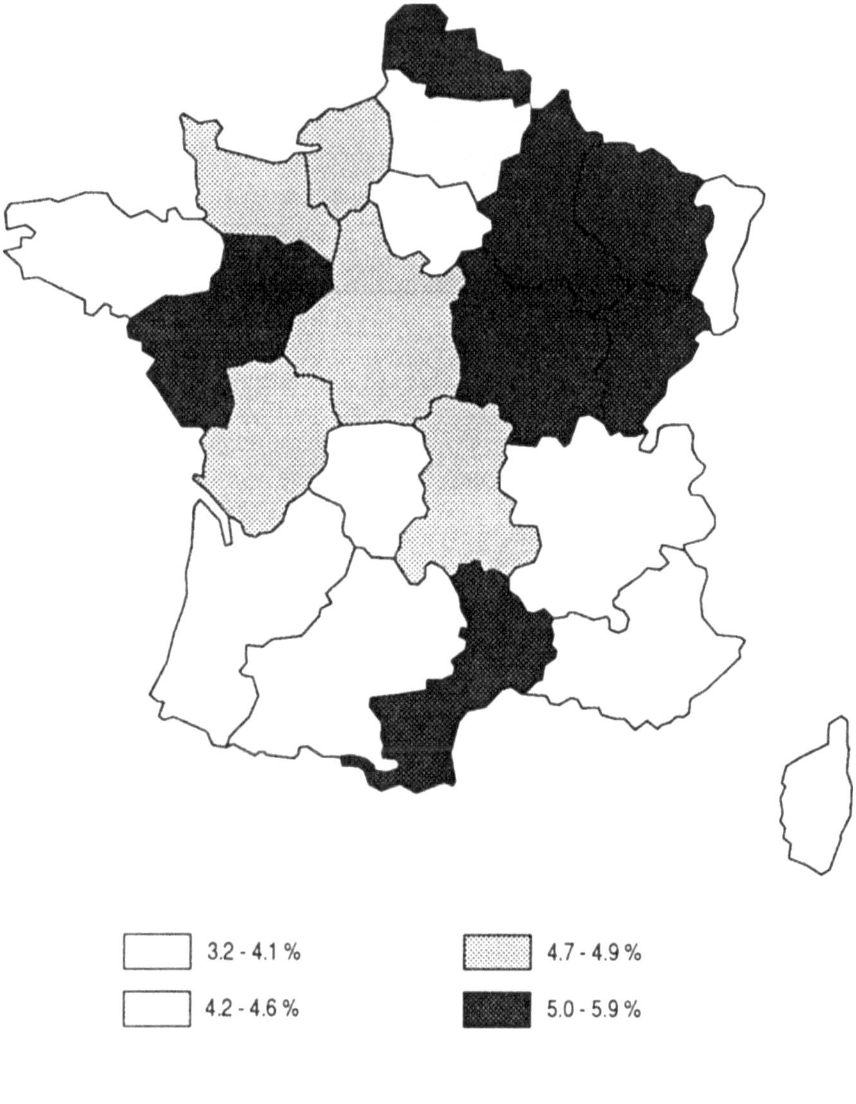

The undeniable process of integration is certainly going in that direction, and the dissemination of production methods and ways of life, which go hand in hand with it, emphasize such a development. However, the structural adjustments it demands produce opposite effects - at least momentarily - which should not be neglected, i.e. the current flaws in economic and social integration. These stress points linked basically to the way the labour markets operate may produce actual splits in the European territory.

Table 5a: Average unemployment rates in 1975, 1981 and 1987, and change 1974-1981 and 1981/1986

Regions	Average unemployment rate in 1975	Average unemployment rate in 1981	Average unemployment rate in 1987	Change in unemployment rate from 1975–1981	Change in unemployment rate from 1981–1987
Ile-de-France	4.4	6.3	8.8	1.9	2.5
Champagne- Ardenne	3.5	8.0	11.4	4.7	3.4
Picardie	3.9	8.6	11.5	4.7	2.9
Haute-Normandie	4.0	9.8	13.0	5.8	3.2
Centre	3.1	6.2	9.7	3.1	3.5
Basse-Normandie	3.7	8.4	11.5	4.7	3.1
Bourgogne	3.0	6.7	10.2	3.7	3.5
Nord – Pas-de-Calais	4.7	10.1	14.1	5.4	4.0
Lorraine	3.1	7.3	10.9	4.2	3.6
Alsace	2.9	5.0	7.5	2.1	2.5
Franche-Comté	3.1	6.9	10.1	3.8	3.2
Pays de la Loire	3.6	7.9	11.6	4.3	3.7
Bretagne	3.8	7.6	11.0	3.8	3.4
Poitou – Charentes	4.1	8.4	11.6	4.3	3.2
Aquitaine	4.4	8.0	11.5	3.6	3.5
Midi – Pyrénées	4.9	7.5	9.7	2.6	2.2
Limousin	3.3	5.3	8.9	2.0	3.6
Rhône – Alpes	3.1	6.2	8.8	3.1	2.6
Auvergne	3.4	6.9	10.4	3.5	3.5
Langedoc-Rossilion	5.8	9.1	14.0	3.3	4.9
Provence – Alpes – Cotê d'Azur	5.6	9.1	12.3	3.5	3.2
Corse	4.2	7.9	11.8	3.7	3.9
France	4.0	7.4	10.6	3.4	3.2

Table 5b: Percentage of long-term unemployed among job seekers at the end of the first quarters of 1986 and 1988, and change therein

Regions	Percentage of long-term unemployed among job seekers at 31/03/86	Percentage of long-term unemployed among job seekers at 31/03/88	Change in percentage of long-term unemployed between '86 and '88
Ile-de-France	26.8	29.6	2.8
Champagne- Ardenne	35.7	32.9	-2.8
Picardie	32.2	31.6	-0.6
Haute-Normandie	33.9	34.8	0.9
Centre	29.9	30.9	1.0
Basse-Normandie	29.3	34.1	4.8
Bourgogne	33.7	33.7	0.0
Nord – Pas-de-Calais	32.7	33.3	0.6
Lorraine	30.3	30.1	-0.2
Alsace	26.9	27.2	0.3
Franche-Comté	31.9	31.9	0.0
Pays de la Loire	33.9	35.4	1.5
Bretagne	31.9	34.9	3.0
Poitou – Charentes	36.0	34.7	-1.3
Aquitaine	29.3	30.9	1.6
Midi – Pyrénées	29.1	29.3	0.2
Limousin	32.3	31.5	-0.8
Rhône – Alpes	28.4	28.0	-0.4
Auvergne	31.2	33.5	2.3
Langedoc-Rossilion	25.6	30.0	4.4
Provence – Alpes – Cotê d'Azur	28.6	27.0	-1.6
Corse	28.8	28.4	-0.4
France	30.2	31.1	0.9

Figure 7a: Development of disparities between member states, per capita GDP
Source: EEC 3rd Regular Report

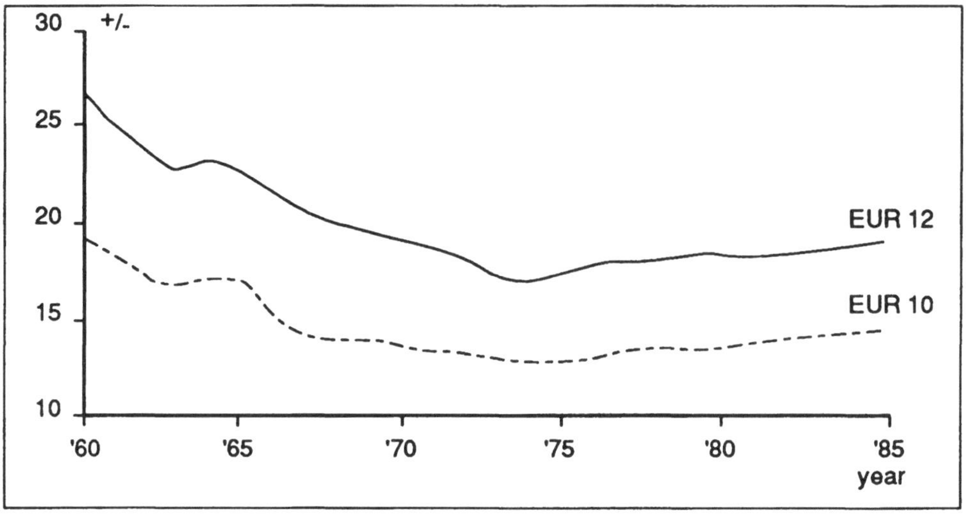

Figure 7b: Development of disparities between member states, GDP per person employed
Source: EEC 3rd Regular Report

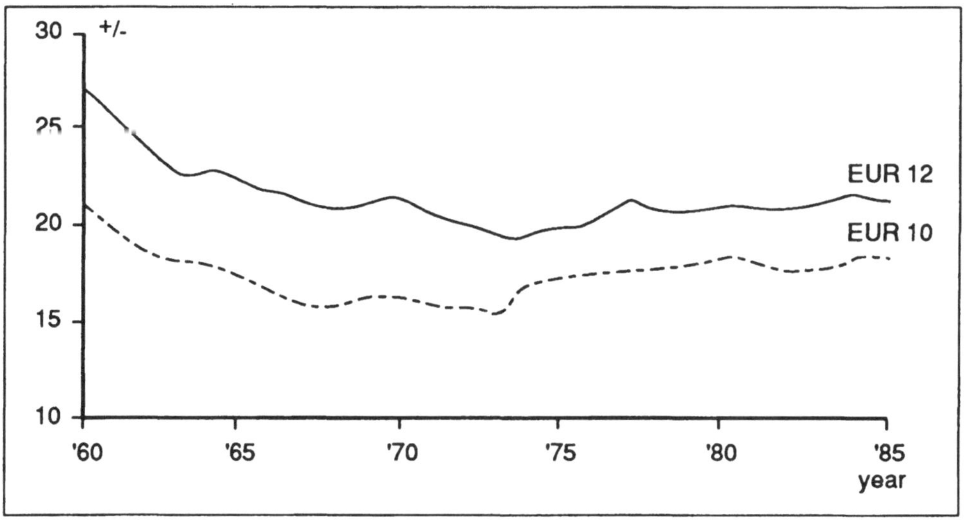

4. A Reversal in Trends?

In recent years the theory of a reversal in trends has been developed which, faced with a decline in the old industrialized regions, sees a new generation of region establishing itself which passes without a transition phase from the first to the third industrial revolution, i.e. from a predominantly agricultural economy to a post-industrial type tertiary system. This idea is particularly strong in the work of Keeble and his team at the University of Cambridge (Keeble, 1983). In the 1970s the peripheral regions of southern Europe experienced greater job growth, as if being rural and agricultural for so long had changed radically in a crisis situation into a substantial development asset, or at least into a situation more favourable to structural adjustment. Herein lies the origin of the common, superficial idea that new growth should result from this southern response, that the future is to be found in the south. Many regions correctly located according to this new "theory of climates" have shrewdly turned it into a promotion slogan, similar to the American hype for the "Sun Belt".

Aside from the magazine articles, there is of course a real phenomenon which is amply reflected in the statistics. This phenomenon cannot be explained by the attraction of sunny shores, but by a new type of upsurge of these regions which, never having experienced the second industrial revolution (the big coal, steel and metallurgical industries), are now free from any major conversion constraints to use their abundant labour force (often highly qualified thanks to the judicious decentralization of schools and universities) for activities immediately adapted to the post-crisis economic conditions: competitiveness, outward- lookingness, new technologies, etc.

In the final analysis, the southern response is but the media's reflection of a reality which is the counterpart of what happened in the industrial revolutions in northern Europe, owing to the close proximity of major deposits of coal and iron ore. Furthermore, in a country like France, this southern response is at the same time and for the same reasons a western response. Between 1975 and 1984, the highest average GDP growth rates were found in the western and southern regions, i.e. right in that part of France which regional policy had made a priority area because it had hitherto lagged behind in terms of industrialization and indeed economic progress (cf. Fig. 8).

The phenomenon also seems of particular note in the Federal Republic of Germany, where if one is to give credence to a study conducted by an institute in Rhineland-Westphalia, the respective losses in the northern and southern regions in terms of national GDP have for more than 20 years shown scissor curves; the predominance of the south becomes apparent in the early 1970s, i.e. at the time of the first noticeable impact of the crisis (see Fig. 9).

Figure 8: Rates of average annual regional growth from 1975 to 1984
Source: INSEE France

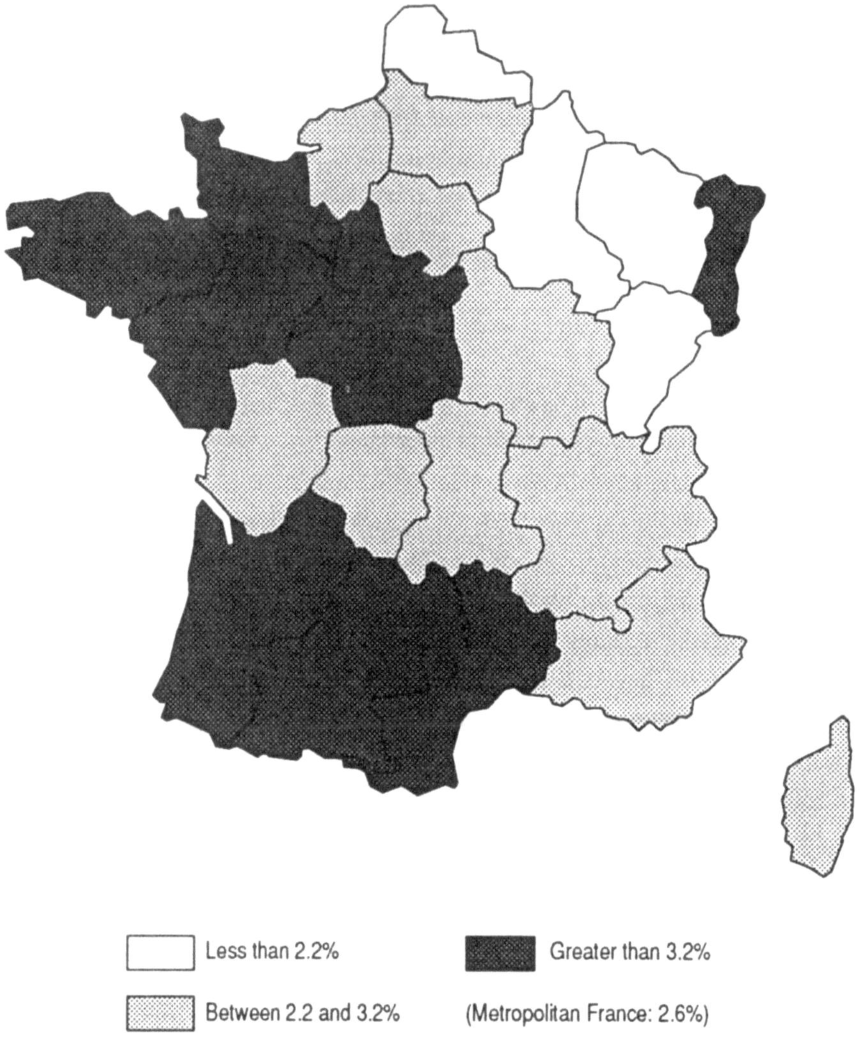

A French study is dedicated for the most part to analysing this phenomenon (Uhrich 1987). Its title, The opposite France?, bears witness to this, even if the question mark emphasizes the inevitable difficulties of the analysis. Apart from the elements already noted, Uhrich's model is based fundamentally on the idea that since these regions have not undergone the strain of industrial conversion they have the best conditions for investing in the fields which appear nowadays to promise success: communications, research

Figure 9: Evolution in the respective share of the northern and southern regions in the gross domestic product of FRG (in %)

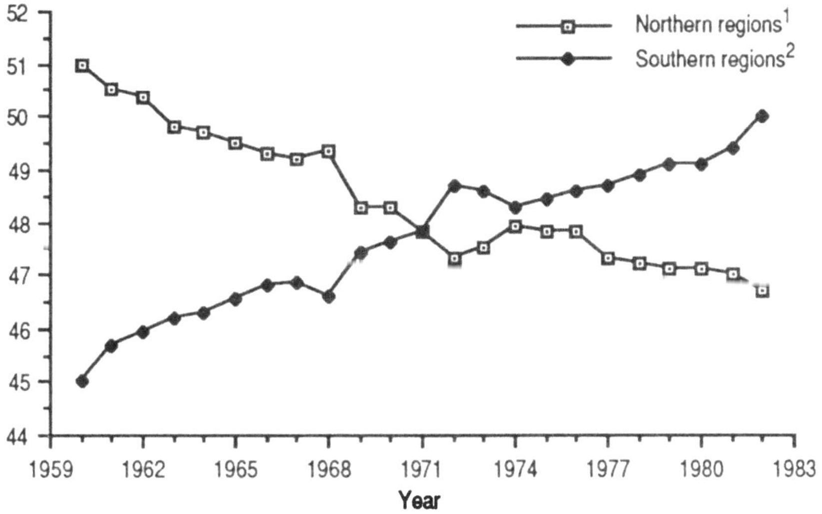

1 Schleswig-Holstein, Lower Saxony, Hamburg, Bremen, North Rhine Westphalia
2 Hessen, Rhineland-Palatinate, Saarland, Baden-Württemberg, Bavaria

Source: Institute of Economic Studies of Rhineland-Westphalia. Revue calculations. Quoted by Uhrich

and technology transfer, training, opening to Europe. The results of the last decade in these fields do not contradict this analysis for these regions have managed to attract a growing share of highly qualified people, researchers and university academics; they also appear at the centre of a real dynamism to create businesses and jobs (see Fig. 10a and 10b).

This leads to a typology of the French regions which in some way reflects a reversal: the regions of the formerly fertile crescent of the northeast are now losing out to the regions of the west and south coasts, while the rest of the country is divided into the "river" regions, including the capital, and the "dry diagonal" leading from the northeast to the southwest, which is in the process of "becoming a desert" (see Fig. 11).

The theory of a reversal of trends seems therefore to be well-founded in part. On condition that it is not confused with a simple attraction to the sun, it highlights the conditions which are likely to enable the regions to overcome the structural and economic difficulties of the decade. It therefore emphasizes the possibility of a new dynamism in regional

Figure 10a and 10b: The renewal of establishments by region (Source: DER-INSEE)

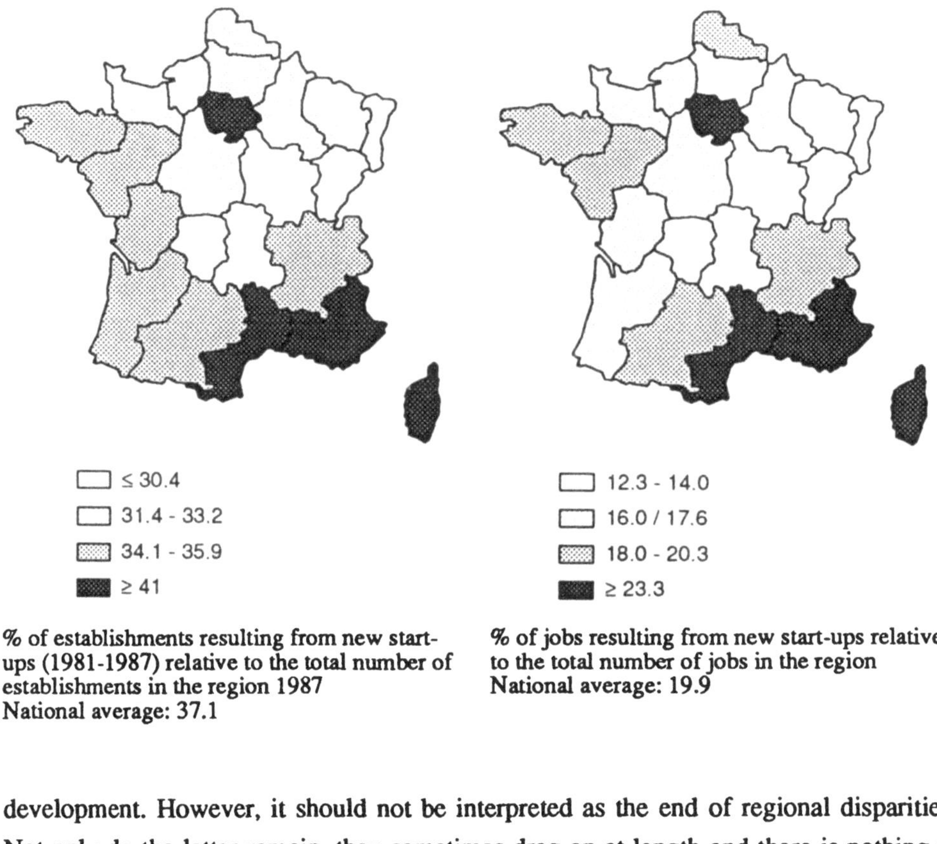

% of establishments resulting from new start-ups (1981-1987) relative to the total number of establishments in the region 1987
National average: 37.1

% of jobs resulting from new start-ups relative to the total number of jobs in the region
National average: 19.9

development. However, it should not be interpreted as the end of regional disparities. Not only do the latter remain, they sometimes drag on at length and there is nothing to rule out the theory that they might grow; for this new dynamism also benefits some regions which are already strong and attractive.

This is especially the case in France with the Paris region and to a lesser degree Rhône Alpes or Alsace (the "strong river regions"). The quantitative rebalances produced by 30 years of dynamic regional policy and the decentralization of jobs are probably more than compensated for today by a qualitative concentration which gives rise to new imbalances, especially, for example, in the spatial spread of public and private sector researchers (more than 60% are located in the Paris region) in precisely those sectors which appear decisive for the future.

France is not necessarily the only case, although in Europe it would probably receive the prize for the concentration of gray matter. Other countries show comparable physiog-

Table 6: Creation of new establishments 1981-1987
Source: INSEE - France

	Jobs %		Establishment %	
	New start-ups	Take-overs	New start-ups	Take-overs
Ile-de-France	25.8	7.8	43.6	10.2
Champagne- Ardenne	16.0	10.5	33.2	12.9
Picardie	12.9	14.0	32.2	14.3
Haute-Normandie	17.1	14.0	32.5	17.0
Centre	17.6	11.5	32.9	12.4
Basse-Normandie	16.5	13.8	32.4	16.4
Bourgogne	13.7	14.6	30.0	15.4
Nord – Pas-de-Calais	18.0	14.8	33.0	15.2
Lorraine	14.1	15.0	31.5	15.3
Alsace	14.0	11.5	32.5	15.4
Franche-Comté	12.7	9.2	30.4	14.1
Pays de la Loire	18.7	11.8	34.1	14.3
Bretagne	20.3	11.6	35.0	14.4
Poitou – Charentes	16.2	11.7	34.5	11.7
Aquitaine	17.6	10.3	35.8	12.5
Midi – Pyrénées	19.6	11.2	35.1	11.9
Limousin	13.9	13.2	28.4	12.9
Rhône – Alpes	19.0	11.9	35.9	13.0
Auvergne	14.0	9.8	29.2	13.7
Langedoc-Rossilion	23.3	8.3	42.4	10.7
Provence – Alpes – Cotê d'Azur	23.5	11.7	42.8	13.7
Corse	25.5	14.6	41.0	14.0
France	**19.9**	**11.0**	**37.1**	**12.9**

nomies in varying degrees (e.g. Spain with Madrid and Barcelona or the United Kingdom with London) and for this reason it would seem that the question should be asked of the whole of the European Community. The single market will probably in fact strengthen the trends towards concentration, with competitiveness resulting from economies of scale. As this process continues how is the gradual forming of a "segmented" Europe to be avoided, i.e. made up of strong areas around a few cities separated by areas which are becoming deserted? While the Commission is already attempting to outline a reply by way of a future reform of the structural funds, a decision which was adopted at the European Summit in 1988, it is obvious that this reply will not be enough if it is not backed up by efforts undertaken by each member state to avoid the emergence of new qualitative disparities.

5. Conclusion

I would like to conclude by expressing a conviction: territorial disparities raise a problem of economic policy since they measure divergences and therefore inequalities vis-à-

Figure 11: The new spatial reality
Source: R. Uhrich op. cit.

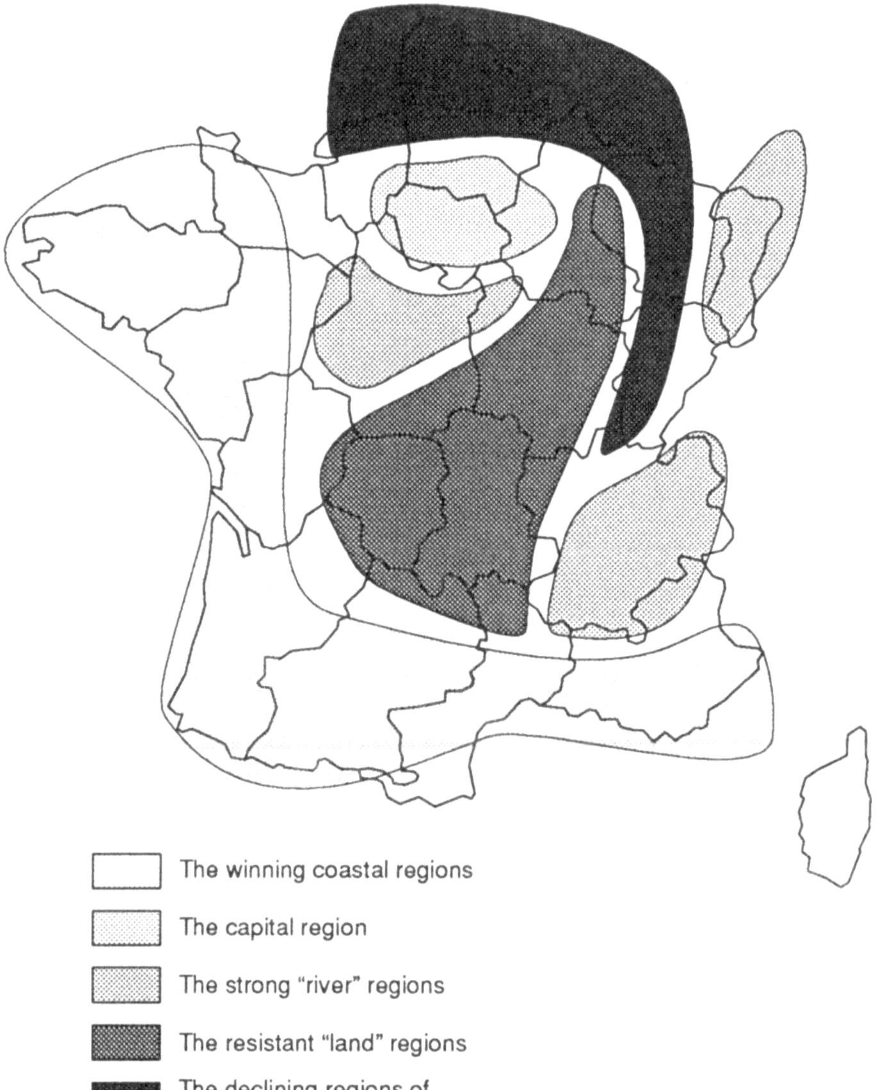

vis a same standard of development; in a positive sense they could also signify strategies of differentiation. While it is in fact illusory to think that a real balance can be achieved by planing down the disparities, it is reasonable, on the other hand, to believe that it is possible for each region or territory to exploit fully all its development assets.

While true inequality is that which restricts or even represses, equality is to be sought in the chances given to regions and territories or in those which the territories and regions give themselves. The consequence which follows on logically from this concept is that territorial development must not be locked up in the rigid brief of a portfolio of activities, awarded once and for always and seen to represent the economic "vocation" of a given region or territory. The steel industry vocation of Lorraine or the mining vocation of Nord-Pas-de-Calais or the Ruhr have sense only in historical terms stretching over one short century; they have hardly any sense in broad historical terms affecting these same territories. There is room, therefore, for a certain "flexibility" of regional productive systems, which no more means volatileness of activity than flexibility of labour means instability of employment. It implies a constant adaptability to the sources of productivity, and, as a consequence, to the restrictions of a market economy. This adaptability is obviously not possible for most regions in every sector. It well and truly assumes a differentiation strategy, which itself evolves. It assumes especially that investments are given priority in those areas which are able to increase the qualitative potential of the territory, notably in terms of research, technology transfer, training, etc.

The lessons of the current decade are conclusive in this respect as has been emphasised above. A study carried out by The Netherlands Economic Institute (Uhrich 1987) in 1982 provides an additional argument. This study provided for a classification of European regions according to their "accessibility" and "receptiveness" to ideas and innovative investments (accessibility being defined basically in terms of qualification potential and receptiveness in terms of receptive infrastructure and level of investment). The classification which results from this study is obviously very instructive. At the top, i.e., with the most favourable assets both in terms of accessibility and receptiveness, are the regions which are either at the top of the classification undertaken by the EEC with the help of its synthetic indicator (see above) in the last two "regular reports", or they are among the regions which have witnessed a reversal in trends mentioned above, i.e. those regions involved in a very dynamic growth process (see Table 7).

Accessibility and receptiveness are therefore presumed to be important factors in regional development since they seem to characteryse just as much the already wealthy regions (there is hardly any change at the top of the list in the last two regular reports at the end of the 1970s and the mid-1980s) as those which are well on their way to prosperity. For this reason almost all regional policies aim to encourage the development of these assets (OECD 1986).

Table 7: Synthesis of accessibility and receptiveness to innovative ideas and investment in the European regions. (In each group the regions have been classified in a decreasing order of accessibility and receptiveness)
Source: NEI Uhrich

RECEPTIVITE	Accesibility Very low	Low	Average	High	Very high
Very low	Tibas Adj. Algarve Alentejo Extremadura Leon Ionioi Nisoi Nisoi aig. Thraki	Galicia Asturias Castila la M. Moise`			
Low	Iles Baleares Iles Canares Peloponnesos Epeiros Thessalia Kriti	Andalousie Murein Pais Valencino Aragon Navarra Castila la M. Norte Centro Sicilia Calabria Basilicata Pugla Marche Abriyyi Trentino Macedonia	North Ireland East Midlands Noord/Nederland Vlaanderen Wallonie Limousin Bourgogne Centre Bretagne Basse Normandie Picardie Champagne Veneto Toscane Campagna		
Average		Fyn Valle d'Aosta Friuli Umbria Sardegna	Lisboa Midi-Pyrénées Aquitaine Poitou/Charentes Auvergne Pays de Loire Haute Normandie Franche–Comté Lorraine Nord Jylland Oost-Nederland South West Wales East Anglia Yorckshire North Piemonte Emilia Romana Liguria	Zuid Rhône-Alpes	
High		Pais Vasco	Cataluna Schleswig-Holstein Rheinland-Pfalz Saarland Luxemburg Langedoc-R. Provence Corse Scotland Ireland	North West West Midlands Sjælland Berlin Niedersachsen Bremen Bayern Baden-Württemberg Alsace Lombardia	
Very high				Sterea Hellas Madrid	Hamburg Lazio Brabant West-Nederland Nordrhein-Westfalen Ile de France South East

Such an option is convincing. It nevertheless overshadows the other major condition in the positive development of territorial disparities: how not to overlook that accessibility and receptiveness are the assets of large cities such as Paris, London, Amsterdam, Bonn, Munich, Hamburg, Madrid, Milan, etc. How is too great a concentration of the cities to be avoided? As noted above, this question calls for answers which, for the main part, can come only from the countries in question and from their policies, be they concerted or not.

Over and above its many macroeconomic virtues, the European single market will also be extraordinarily revealing of the chances of each region to affirm its specific identity or, on the other hand, of the derisory nature of its efforts in the face of a poorly managed recentralization of power and responsibilities. More than ever before regional policies will have to choose between these two scenarios.

References

Allen, K. et al. (1988)
European Regional Incentives. European Policies Research Centre, Glasgow.

Aydalot, P. (ed.) (1984)
Crise et espace. Economica, Paris.

Aydalot, P. (1985)
Economie spatiale et urbaine. Economica, Paris.

CEE (1984)
Deuxième rapport périodique sur la situation et l'evolution socio-économiques des région de la Communauté. Bruxelles.

Cuadrado, R. J. (ed.) (1987)
Los desequilibrios regionales y el estado de las Autonomias. Ed. Orbis, Barcelona.

INSEE (1984)
Statistiques et Indicateurs des Régions Françaises Annese projet Loi de Finances 1985. Paris.

Keeble, D. et al. (1983)
Updating of centrality - peripherality and EEC regional development study. Cambridge.

OCDE (1986)
L'évolution des politiques régionales dans les pays de l'OCDE. Paris.

Martin, R. and B. Rowthorn (1986)
The geography of deindustrialization. Macmillan, London.

Urich, R. (1987)
La France invers? Les régions en mutation. Economica, Paris.

3 Regional Development Potentials and Strategies for Strengthening Regional Innovation Capacities

3.1 Regional Development Potentials and Innovation Capacities

Peter M. Townroe

1 Introduction

Until the 1980s, technology and innovation were under recognised influences in the explanation of differences in the rates of economic growth between regions in advanced industrial nations. As a supply side factor, technology is disguised within statistics on investment and capital stock. Most nations do not have adequate statistical series which directly reflect technological capacity or performance at the national level, let alone with a regional disaggregation. Only in the past decade have proxy statistics and data from focussed surveys been brought to bear. This new work has pushed the technological aspects of regional economic development back onto the centre stage for both theoretical analysis and for policy discussion. This is reflected in recent literature (e.g. Maillat, 1982; Thwaites and Oakey, 1985; Nijkamp, 1986; Howells, 1987; Howells and Green, 1987; Keebel and Wever, 1987; Giaoutzi, Nijkamp and Storey, 1988).

Following this introduction, Section 2, is a brief discussion of the nature of technology, making some important distinctions that will prove useful in considering the role of technical change in regional economic development. Particularly relevant here will be a recognition of a number of contexts which influence the technological capacity of a regional economy.

Section 3 briefly discusses the role of technology in the regional economic development process, noting a number of "pathways" by which the regional economy can benefit from technological advance, and speculating as to why in recent years innovation has been receiving a new premium in dicussions on both national and regional economic growth.

The case for a policy intervention to support technical change at the regional level is taken up in Section 4. Some of the difficulties in effectively delivering appropriate policy instruments are noted, without a detailed consideration of what these various instruments might be, nor drawing upon the policy experience of any particular nation. This paper is deliberately kept general. The self-appointed task here is to raise issues and to pose problems: not to prescribe or to evaluate policy measures. It suggests ways of thinking about innovation and about the part an innovation focussed element may play in a regional economic development policy.

At the outset, it should be recognised that there are two perspectives which might be taken in examining technology and innovation within the context of regional economic development. The first could view the delivery of policy measures designed to support technical change in industry and commerce at the local or regional level as a mechanism of support to accelerate economic growth in the overall national economy. It is a national view. The second perspective is more localised. It considers policy contributions to enhanced innovation and adoption of new technology in an economically lagging region as one of the policy options for reducing regional inequality in economic performance within a national economy. This paper takes the second perspective, but will return to the first in the Conclusions.

2 Characterising Technology

"Technology" is a complex entity to pin down in an economic sense."Technical progress" is controversially seen in a time trend term, or in the "Hicks Shifter" constant, of an industry specific or economy wide production function. The temptation is therefore to see technology sitting in alongside labour and capital as a third factor of production. If the state wishes to see a given industry or a given region increase its levels of output, then, by this view, all the state has to do is to find ways and means of tipping in more technology. The technology joins the other factors, rather like a fuel additive, and up goes the rate of output.

In fact, of course, this view is far too simplistic. Technology is not just a third factor of production, with relevant cross elasticities to the other factors and an elasticity with respect to output. Technology is bound up within the stocks of capital and labour available to a regional economy. It is part of the profile and characterisation of the machinery and equipment and the industrial and commerical buildings in the economy, as well as of the infrastructures and of the education, skills and experience of the workforce. It determines not only the capacity of stocks of capital and labour to produce valued output, but also (and most importantly) the capacity of those stocks to be both flexible in the face of exogenous changes in demands for the goods and services produced by the region and expansive in generating additional capacity for future production. The present technology contained within the labour and capital of the region will determine the pace of future technological development within local industry and commerce.

Technology is necessarily involved in both the products of a regional economy and the processes by which they are produced, in both the goods sectors of the economy and in the service sectors. The evolution of a given technology, or its trajectory, is therefore closely related to the life cycle of the product(s) which it serves; where this "serving" comes about from both improvements in the means of production and in the development of the product specification. The progressive application of new techology is therefore seen in both cost and quality and also in the variety of the product. Technology may be embodied within the machinery and equipment and the buildings that are used; or it maybe disembodied, being seen in the procedures, the forms of organization, the ways of working: it is hard and it is soft. It will interact with market demand to determine the life cycle of each product. It will also interact with investment and profitability to determine the history of each firm.

The theory of contestable markets encourages us to see technology as one of the (many possible) barriers to entry for a firm into a given sector, and possibly as a barrier to exit as well, even though innovation is commonly seen as a factor facilitating entry. Costs incurred in developing new technology are therefore 'sunk' costs and may constitute a barrier to entry by a local company into a new product area or by an external company into a product area fostered by a local company. Sunk costs are different from 'fixed costs', where the barrier to entry effect is small if there is a resale value to the asset, or it can be transferred to an alternative use. This sunk cost/fixed cost distinction highlights an important public policy choice: should the state foster indigenous innovatory capacity or should it encourage the external adoption and purchase of the latest "technological frontier" technology (normally through the purchase of capital equipment but also through the purchase of specialist design and engineering services)? Or is it appropriate to seek to distinguish, for policy purposes, the process of innovation from the process of adoption? Further, on this theme, the human capital tied into the skills and the knowledge of the workforce may be regarded as a sunk cost if the education and training costs incurred for a specific industry or technology cannot be retrieved and applied to other outputs and other processes. This says something about the importance of basic or primary knowledge in the education process and of training for flexibility in the application of skills and know-how. Both innovation and adoption of new technology require a move away from a "once-off" view of education and training. The benefits of much modern technology come from a flexible workforce, with diagnostic skills, with an ability to learn and to assimilate new technical information, and (ideally) with an ability to apply the new technology creatively. This may be regarded as a call for entrepreneurship at the work bench.

Technology may be further characterised as "releasing the potential" tied into the existing capital and the present portfolio of labour skills of the regional economy: its economic endowment. This is accomplished by improving the productive use made of the stock of these resources. Or technology may be seen as "generating capability" in the economy: moving the economy into new directions in product and process, where local invention and innovation or the adoption of new technology from elsewhere enables the regional economy to reorientate and to accept further technological development. Technology here requires investment, in new capital equipment and in new labour skills. The development of the necessary scientific and engineering skills and of understanding and application by production workers will be the key to the commercial success of both of these inputs of new technology.

It needs to be remembered that both the acceptance of new technology and the productivity of the expenditure incurred in that acceptance will be directly influenced by a number of contexts. Each context must be brought to bear on any discussion of regional technology policy, or of regional development measures (subsidies to investment etc.) which have a technological component.

The most general context is the most diffuse. It is tied into local culture and social attitudes and changes only slowly. It may be termed "acceptance of progress". Not everyone sees change, certainly all change, as progress. Within a national economy some regions may have an inheritance of resistance to the introduction of new technologies. This resistance has often been well founded, when a new technology outdates existing skills, forcing workers into unwelcome retraining or the acceptance of second best (as they see it) employment. But such resistance can develop attitudes which retard investment in research and development as well as in retraining, and maintain the (mis)understanding that education is only an early life experience, with little relevance to changing circumstances thereafter.

This context of attitude clearly influences the ability of a regional economy to react to the changing technological strength of other regions and other nations and to foresee that a present trading strength of industries within the region may decline unless emerging competitors can be matched, on price and quality. The issue here for policy is whether the local region in question can realistically hope to meet the technological and production challenge that is coming up over the horizon for all of its present industrial sectors. Should policy be selective in the areas of technology to be supported?

The forms and dispositions of the financial agencies (banks, investment houses, stockbrokers, insurance companies, etc.) upon which the industrial and commercial companies in our region are reliant provide a further context for technical progress. Attitudes towards supporting start-up capital differ, as do attitudes towards lending for non-asset based expenditures like research and engineering oriented development. A gap here may point to a role for a state agency or a regional development fund or an adjustment in tax treatment for these kinds of expenditures.

In some areas of technology the local context for the individual firm of sources of advice and interpretation of the science and engineering involved may also be important. This ties in locally based consultants, as well as local universities and colleges. Access to specialist subcontractors will also be relevant. Many of the current advanced machining centres and the associated advanced company wide information systems are allowing companies to develop economies of scope: cost savings from the joint production of differentiated products. But such differentiation may require a wider variety of external contact and support for the firm than hitherto. Finally, an awareness has emerged from researchers and commentators over the past decade that technological change in the advanced industrial nations has been influencing the organizational structure of industrial and commercial companies and the nature of the labour process within the companies. The key to these changes has been the progressive adoption of advanced information technologies at all levels and stages of production. The evolving paradigm, of what has come to be called the "post-Fordist mode", is of a very flexible organizational structure, loosening up bureaucratic forms with the firm, devoted to greater customisation of products and to a more rapid response to consumer demand. This involves a core flexible labour force which is closely identified with the company and then adjustment in the labour input is achieved by the successive recruitment and lay off of groups of peripheral workers. These developments place a new emphasis on technology, and on the generation and acceptance of innovations, both "hard" and "soft".

3 The Technology Factor in Regional Development

In most European nations the provision of infrastructure for industry in each region now follows common standards. The same is true of primary and secondary education. By virtue of national wage agreements, the price of specific skills also does not vary greatly between regions. And the price of capital is similar, with small variations in risk premiums. Given that the transport cost element is so small in the majority of industrial products, the overall cost variations which might influence flows of capital and labour between regions are therefore very small. This is one reason why any disequilibrium in growth rates between regions is expressed primarily through the numbers of unemployed. And yet, looking around Europe, there are very clear geographical differences in innovation activity and in adoption of best practice technologies between different regions and subregions (e.g. Howells 1987).

The overall long run objective of regional economic development policies may be taken as the raising of income per capita, or of real household disposable income. So policy targets such as raising the capital per worker, improving the value added contribution of labour, or accelerating access to available technology may be characterised as implementational objectives rather than strategic objectives. The technology arm of regional economic development, therefore follows an implentational objective. It operates through five different "pathways".

New technology may come to industrial and commercial companies in a region through investment, as the companies purchase new plant and equipment embodying new capabilities. This is by far and away the most common way of acquiring new technology. It is the dominant pathway. Alternatively migrant companies, locating a new branch plant or transferring their operations into the region, provide a second pathway, but for the regional economy as a whole. A third route is through intracompany transfers of skills and know-how between plants but into the region from elsewhere. Any policy of support for each of these pathways necessarily relies on the companies themselves to judge the appropriateness of the technology they acquire and which they then put to use. Any policy influence on the direction of technical change is minimal.

Two further pathways may be identified. The first is the generation of technological advance within firms already located in the region through expenditures on research and development and through ad hoc reactions to customers involving technology and to ideas and innovations generated from the production and marketing processes. Companies may receive direct support for this activity from all levels of government, either in general terms or by focussed research or development contracts. Or the support may be indirect, through subsidy to consultancy or use made of local colleges etc. The second pathway is one which has been receiving increasing attention in recent years: spin-off.

"Spin-off" in this context has two meanings. The first is an externality from one regional company to another, brought about by local information flows (customer-supplier negotiations, discussions within professional and scientific societies, local trade fairs etc.). One company learns from the other. The information exchange here requires an ability to understand and to assimilate. The other meaning of spin-off arises from "outreach" by local universities, colleges and state-funded research institutes. Staff members from these institutions may start up small companies based on specific technological knowledge and skills, either on the side or by resigning their posts. Links back to the institutions may then be important in the early life of these companies. Alternatively, these institutions undertake contract research for regionally based concerns and/or provide analytical services. Or their presence attracts investment by high-tech companies into the region, with a research "listening-post" or laboratory, or a production facility that can directly benefit by proximity to the research centres. Cambridge, England is a well-researched example of this type of spin off. (Segal et al. 1985). Many science park developments across Europe have been launched in the hope of fostering spin-offs of this kind.

All of these pathways, directly or indirectly, seek to raise the productivity of capital and labour within the region, as well as attracting new capital and labour at the margin into the region. The skill base required to maximise the advantage to be taken of each pathway may be characterised as knowledge based, diagnostic, flexible and entrepreneurial. This is in contrast to the more static procedural skills found in many companies which are largely function specific and so in danger of being unresponsive and resistant to change.

4 Policies for Innovation and Adoption

The reluctance of many commercial banks to provide loan finance for research and development expenditure, noted above, might be characterised as an example of market "imperfection" or market "failure". This failure then constitutes a case for public policy intervention, providing a social rationale for action where the private incentive was insufficiently strong. Other grounds for governmental intervention which have specific relevance to a technology policy at the regional level also arise from this market failure perspective. All revolve around an acknowledgement that market forces alone are unlikely to produce the desired level of economic achievement in the regional economy (viz a closing of the gap in regional GDP per head with a national average).

Invention, the discovery of new knowledge, typically has the characteristics of a public good rather than a private good. In other words, it may be consumed/acquired by one economic agent without depriving another economic agent, and it is difficult to exclude anyone from the knowledge except by complex law (patent law, copyright law) or by costly secrecy. The pay-off to research and development expenditures by a single investor is therefore uncertain. Society may benefit but the original developer may not be sure of sufficient benefit to justify his investment. There is a positive externality, under which the social gain exceeds the private gain. For smaller companies in particular the financial resources required to defend through courts a patent that is being violated are typically so large a burden that it is both cheaper and safer not to bother with a patent in the first place. The resources are better used to further advance the technology, paying scientists and engineers rather than lawyers. But overall in this situation, the price system does not provide the right incentives for the interests of society. There is therefore a case for collective expenditure on the invention process, through subsidy to companies from the state, or by support for universities and research institutes. This general case for state action in support of scientific research is well recognised (e.g. Arrow 1962; more generally, Freeman 1982).

Less well recognised is the case for supporting innovation, the incorporation of new knowledge into actual production techniques (see Freeman 1987; Stoneman 1987). Here, market forces should act as an arbiter. These forces decide which new knowledge, either embodied in new machinery or developed into procedures for action by a labour force, it is worth bringing into the production process: which new knowledge results in goods and services upon which the consumer will place a positive value. In fact, however, the process is not so straightforward, particularly in a lagging region.

If a region is lagging in terms of productive achievement relative to its capacity, then that may say something simply about an inadequate shift in the regional economy in the portfolio of goods and services offered for sale against recent changes in taste or against new competition from non-local companies for sales both within the region and elsewhere. Over time however, one might expect this shift to take place, as resources of land, labour and capital in the region get reassigned.

Alternatively, it may be that the regional economy is lagging in terms of a productive potential, rather than in the use of capacity. "Potential" is taken here to mean a level of capacity in excess of that currently existing, given the industry mix of the region, where the potential capacity is defined either by a national norm of output per worker or by a "best practice given the available technology" norm of output per worker. In either case the regional economy requires investment in both physical capital and in human capital to achieve that potential, and the introduction of technology new to the industries of the region is necessarily involved. Public policy may support the investment required, but it will be more effective, the argument goes, if it supports the recognition and acceptance of the new technology as well.

The case for state support for technological diffusion also rests on an argument of imperfect information. Here smaller companies in particular are seen as not having the time or the resources to keep up with the technology relevant to their products or processes. This is an economies of scale argument. The case is also sustained by the argument that the private investment premium on the risks associated with the adoption of new technology is unacceptably high from a social rather than a private perspective. Some form of subsidy is therefore justified. It may also be that local, monopoly power will reduce the incentive to keep up with a technological race.

Support for new technology might also be canvassed on a simple time discount argument: that the social rate of time discount is lower than the private rate of time discount. This view sees regional development policy as involved in a long term strategy of building up technical competence, with payoffs that stretch beyond a specific process or product in a specific company, and so beyond a private pay-off. The "sustainability" of the regional economy is involved: sustainable in terms of growth and response to exogenous forces of change. This requires the regional economy to be able to continue to assimilate and commercially apply new technology, creatively adapt what is on offer elsewhere as well as to use new ideas "off the shelf".

Any policy of state assistance to private sector companies which is directed at accelerating the processes of invention and the rates of adoption of new techology involves providing those companies with additional resources: finance capital, trained manpower, knowledge, or advice. Unless, of course, the policy is to proceed by exhortation alone. And it is true that a clever policy will seek to lever out additional private sector resources to match or exceed the taxpayer supported input. Provision of the additional resources at a regional rather than a national level provides the opportunity to tailor policy packages more closely to the needs of the local economy and to enhance coordination between the different arms of government that might be involved. But as with all policy areas of government there are a number of standard issues which have to be faced. There are no easy answers for any of these.

These "policy delivery" issues include:

1. The level of funding, overall and per recipient
2. The nature and control of the agency providing the funding and other relevant services (advice, consultancy, etc.)
3. The incentives required to induce participation by those agencies in the region which see their primary tasks as lying elsewhere (e.g. universities, state research institutes)
4. The supply of skilled manpower at all levels to make the policy work
5. The possibility of a direct social reclaim on the social investments which have been made (via royalty or licensing payments, or contract research profits)
6. The criteria to be adopted in monitoring the policy and in evaluating the effectiveness of the expenditure mix adopted

For anyone involved in studying or implementing policies for technology it is obvious that there are no clear ex-ante answers to the questions lying behind the issues listed. The argument for monitored experimentation in policy design, and for cross-regional and cross-national learning is very strong.

Conclusion

Any region in an advanced industrial country which can no longer rely on the exploitation of a natural resource that is rising in value for the cornerstone to its prosperity - and this means the vast majority of regions - has to engage in active competition, for exports to other regions and against imported goods and services. It is true that competitive failure is in part disguised by the many processes of fiscal transfer that occur between regions in the modern nation state. But over time, failure to bring in new technology to both product and process must condemn the region to a relative fall back in levels of economic prosperity. The capacity to absorb and to take commercial advantage of new technology is an essential part of economic growth. And if a well grounded competitive advantage is sought, it is then secured by going beyond the absorption of the technology of others to a capacity to generate innovations within the region.

The successful delivery of a policy package which results in rates of invention and innovation and adoption in a region faster than would have otherwise been the case necessarily results in a contribution to a faster rate of national economic growth as well as regional economic growth. The national potential is raised. Public money has not been used merely to secure a spatial redistribution of the national assets of labour and physical capital. More than a zero-sum game is being played. So a case can be made for policies of regional delivery of technological support to companies as being in the national economic interest, not solely for regional "catch-up".

Overall regional innovation strategies have to be more than just technology transfer policies. They have to embody elements of support for basic research. They have to involve education and training of both technologists and those who can ensure that the new technology is turned into commodities which consumers will wish to buy. They have to be concerned with the diffusion of scientific and engineering know-how and expertise. And a regional innovation strategy has to both maximise the advantage that is taken of existing innovation capacities as well as raising the potential for the generation and application of scientific knowledge over time. The state must be both an enabling agency and a catalyst.

References

Arrow, K. (1962)
The Economic Implications of Learning by Doing, Review of Economic Studies. 29: 155-173.

Freeman, C. (1982)
The Economics of Industrial Innovation. Frances Pinter, London.

Giaoutzi, M., P. Nijkamp and D. Storey (1988)
Small and Medium Sized Enterprises and Regional Development. Croom Helm, London.

Howells, J. (1987)
Economic, Technological and Locational Trends in Services: The Role of Information Services in European Regional Development. Gower, Aldershot.

Howells, J. and A. Green (1987)
Technological Innovation, Structural Change and Location in the U.K. Services: National and Regional Economic Development Prospects. Gower, Aldershot.

Keeble, D, E. and E. Wever (eds.) (1987)
New Firms and Regional Development in Europe. Croom Helm, London.

Maillat, D. (ed.) (1982)
Technology: A Key Factor for Regional Development. Georgi Publishing, Saint-Saphorin.

Nijkamp, P. (ed.) (1986)
Technological Change, Employment and Spatial Dynamics. Springer, Berlin.

Segal, N., R. Quince and W.Wicksteed (1985)
The Cambridge Phenomenon. Cambridge.

Stoneman, P. (1987)
The Economic Analysis of Technology Policy, Clarendon Press, Oxford.

Thwaites, A.T. and R.P. Oakey (1985)
The Regional Economic Impact of Technological Change. Frances Pinter, London.

3.2 Industrial Innovation and Spatial Systems: The Impact of Producer Services

Evert Jan Davelaar and Peter Nijkamp

1 Introduction

The interdependence between technological change and economic growth is becoming more and more evident. After a period of relative ignorance in which technological change was often seen as a regular and stable "manna from heaven", times seem to have changed; attention has shifted towards the link between technological evolution and economic fluctuations (cf. Freeman et al. 1982; van Duijn 1981; Mensch 1981; Kleinknecht 1987; Vasko 1987). In this framework, the concept of basic innovation (cycles), notably the Kondratief cycle is often proposed as one of the main driving forces for long term economic fluctuations (Mensch 1981). On the other hand, some authors claim that the causal mechanism is just the other way around (cf. Schmookler 1966). Disregarding the issue of whether these "cycles" have a fixed periodicity, two important and as yet unresolved and interrelated topics come to the fore.

First the question can be raised whether such cycles of innovations and economic growth can be empirically observed using time series data. Second, the issue as to which cycle "precedes" the other one - i.e. the identification of the "triggering" mechanism - has to be considered. The assessment of the exact timing of such cycles, and especially the innovation cycle, is however not easy, because of disagreement on both the identification of innovations (i.e. the criteria for selecting different types of innovations) and on the exact timing of the selected innovations. Given the lack of agreement on these two issues (cf. Freeman et al. 1982 versus Mensch 1981 or Kleinknecht 1987), it is clear that the question of the identification of the "triggering" mechanism is still largely unresolved.

This study was supported by The Netherlands Organization for the Advancement of Scientific Research (project no. 450-230-004)

Fortunately, however, one can observe a common acceptance of the existence of "structural adjustment periods", in which economic growth (based on existing industries and markets) stagnates and future economic growth largely depends upon the creation of new markets and industries (cf. van der Zwan 1979; Mensch 1981; Freeman et al. 1982). In this framework, technological transformation is regarded as a prerequisite for further economic growth of these latter sectors. In this respect Freeman et al. (1982) have introduced the concept of "new technology systems". This should be interpreted as a configuration of "bunches" of innovations and related "new" industries that rapidly infiltrate, evolve and become the driving force for a change in the existing economic structure.

Our paper will focus on the space-time patterns of such new technologies and (especially) the related spatial economic transformation. It will start with the presentation of a conceptual-theoretical framework for analysing the space-time trajectory of such new (economic) activities in general. The spatial dimension will be linked to the concept of "technology systems" and the related concept of "innovation cycles". It will then be hypothesized that there is no unambiguous geographical concentration of innovations in a given type of (metropolitan) region. On the contrary, several types of regions are expected to play their own specific role in a "creative" diffusion process related to "new technology systems". The evolution of regional economic performance related to these systems is assumed to mirror the timing of this "creative role".

It is noteworthy that in particular the producer service sector is often considered to be such a dynamic growth sector (cf. Daniels 1987; Noyelle and Stanback 1984; Lambooy and Tates 1983; Bearse 1978; Freeman 1987) which is related to (basic) innovations in the field of information and telecommunication technology. Such a sector may not only be relevant from a national-economic, but also from a mesospatial point of view, as it may offer opportunities for restructuring metropolitan economies which in the past decade have been under severe pressure due to the urban-rural shift of manufacturing employment (cf. Keeble et al. 1983; Keeble and Wever 1986). Consequently, in this paper we will try to test the empirical validity of our theoretical framework by tracing in particular the space-time trajectory of this sector in The Netherlands.

Our paper is organized as follows. In Section 2 our theoretical framework will be sketched. Next, in Section 3 the relative growth of the producer service sector in The Netherlands will be decomposed. In Section 4 we will select various subclasses of the producer service sector which belong to more recent "technology systems". Next, Section 5 will be devoted to a discussion of the characteristics of our empirical data base and of some general empirical findings. Section 6 will then introduce the spatial dimension, while in Section 7 the space-time trajectory of our selected subclasses will be analysed. Finally, Section 8 contains some general conclusions.

2 Space-Time Trajectory of New Technology Systems

In this section two concepts, viz. "new technology systems" and "innovation cycles" will be integrated in studying the role of technological change in shaping and re-arranging spatial structures. The first concept has been proposed by Freeman et al. (1982) as a useful framework for analyzing the interrelationships between technological change and economic development. In this concept especially the swarming of "families" of innovations through new "Schumpeterian" firms is considered to be crucial for an upswing in investment and employment. This "swarming" of innovations is essentially related to further improvements and applications of some (related) basic innovations for both product and process technologies.

The "timing" of this "swarming" effect "may not necessarily occur immediately after a basic innovation although it may do so if other conditions are favourable" (Freeman et al. 1982: 65). These favourable "conditions" may refer to other complementary types of innovations (managerial, organizational, institutional) as well as to the mere socio-institutional and political climate prevailing at the moment[1]. Thus crucial for the "timing" of this swarming effect is "a constellation of circumstances favourable to the exceptionally rapid growth of one or more new industries, each involving the combination of a number of related inventions, innovations and economic and social changes" (Freeman et al. 1982: 70). As the "bunches" of innovations generated during this "swarming" period are mutually related, both in a technical sense (to some basic innovations) and as a result of favourable social and political circumstances, they may be labelled "families of innovations". New technology systems are not only characterized by the emergence of one or more new sectors, but also by the large impact on other sectors of the economy.

A second, and related, concept is the "innovation cycle", as described, among others by Abernathy and Utterback (1978), Sahal (1980) and Rothwell and Zegveld (1985). The empirical validity of this concept has been confirmed among others by Freeman et al. (1982) and Davelaar and Nijkamp (1987a,b,c). According to this notion technological change can be considered as a continuous, though varying, process. In the first phase of the "swarming" process the emphasis will be on changing and generating new "products" or "services" (i.e. product innovations) to be derived from the related basic innovations. During later phases of this "swarming" process, however, the emphasis will shift towards the generation of improved (cost-efficient) production techniques for product variants that have become "standard" by that time (i.e. process innovations). Thus, for a certain technology system the following pattern may be assumed for the related innovation cycles (see Fig. 1):

Figure 1: Technology system and innovation cycles

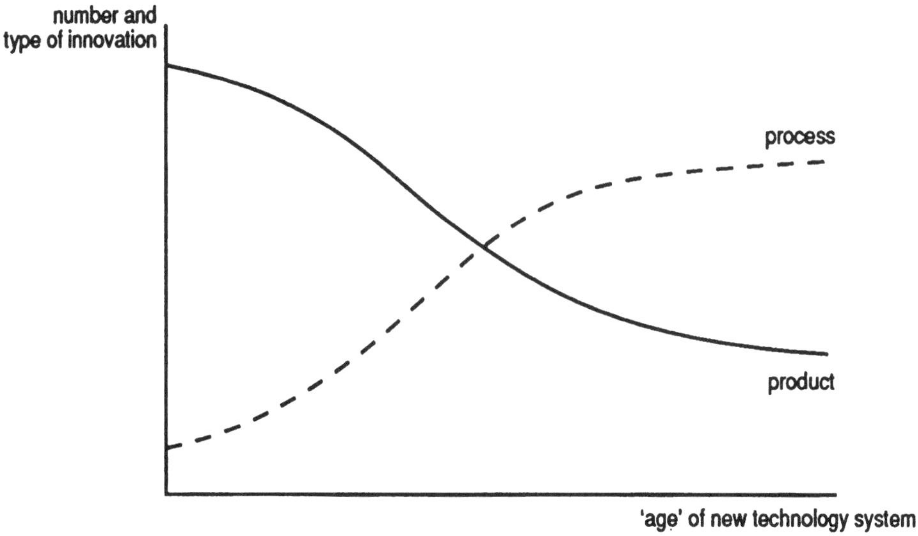

In Fig. 1 innovations are assumed to qualify only if they exceed a prior "quality threshold" (for example, innovations that are new to a whole sector). Essentially, the "innovation cycle" reflects a "learning process", concerning both new products and related production techniques, during the lifetime of a "new technology system". Nelson and Winter (1982) labelled this the "natural trajectory" of new technologies. As a consequence of this learning process, innovation diffusion "cannot be viewed as one of simple replication and carbon-copy imitation, but frequently involves a string of further innovations, small and large, as an increasing number of firms get involved and begin to learn new technology and strive to gain an edge over their competitors" (Freeman et al. 1982:65). The same arguments can also be found in the earlier work of Kuznets (1930), and more recently in Rosenberg (1976), Metcalfe (1981) and Granstrand (1986). Thus the innovation cycle implies a "creative" diffusion process in which the emphasis will shift from product to process technologies. This diffusion process is labelled "creative" in order to distinguish it from those diffusion models and theories which consider the diffusion process to be a static "carbon copy" process. Especially this "creative" aspect of diffusion (and the consequent spatial innovation and related competitiveness patterns) has been largely neglected in spatial innovation theories and models. Unfortunately, most of the, sometimes technically sophisticated, spatial diffusion models consider the spatial diffusion process of a homogeneous innovation, which is static in both time and space. In reality, however, it may be extremely difficult to identify even one such "static" innovation. Furthermore spatial innovation research has largely neglected the interrelations between changing spatial structures, technology systems and "creative diffusion". Therefore, it is the purpose of the present section to make a first step towards integrating these concepts.

In this framework, we suggest the use of a "dynamic" incubation theory in which technology, in relation to both "technology systems" and the "creative diffusion process", acts as a major driving force. In this context, we assume that the following three phases may roughly be distinguished:

The incubation phase. In this phase, which is related to the "take-off" of a "new technology system", the effects in terms of "new" and "innovative firms" are expected to be first noticed in the larger and central metropolitan areas. On the supply side, information flows (cf. Pred 1977), availability of a large and highly skilled labour force (especially relevant at the introductory phase of completely new products (cf. Thompson 1970, Oakey et al. 1980; Howells 1983) and social overhead capital (cf. Hirschman 1958)[2] may favour these areas. On the demand side the availability of a large and diversified market and availability of "opinion leaders" (cf. Berry 1972 and Brown 1981) may be considered as factors favouring metropolitan areas (cf. Malecki and Nijkamp 1988). In this phase metropolitan areas will act as a seedbed for many product (or service) innovations (as can be derived from the pattern depicted in Fig.1) and related new "Schumpeterian" firms.

The catching-up phase. As the "creative diffusion" process along the "natural trajectories" goes on, however, on the supply side products become more and more standardized (e.g. the video VHS system becomes the "standard") and the emphasis shifts from product to process innovations (cf. Fig.1). Information flows (i.e. intensive communication patterns between consumers and producers) and skilled labour become less important locational pull factors (cf. Andersson and Johansson 1984; Batten and Johansson 1987; Thwaites 1978). Concerning the demand side, the markets in metropolitan areas will, as a consequence of being market leaders, first approach the saturation level. Consequently, both supply and demand factors make metropolitan areas less advantageous, and hence (part of) the production will shift[3] to non-metropolitan areas. As will be clear, these regions will reap the fruits of the "learning economies" achieved so far (i.e. the fruits of the "creative diffusion" process) and direct their investments towards the "best"[4] production techniques (process innovations). They will also concentrate on the production of and further (though less radical) adjustments of (new) products that have proven most successful. As is clear from Fig.1, the resulting investments made in the non-metropolitan areas will, compared with the preceding "boom" phase in metropolitan regions, result in a great many process innovations being generated and applied in these areas.

The competition phase. When all consumer markets become saturated and possibilities for further improving the newly generated products or services diminish, price competition becomes more important in increasing a firm"s market share. As a result of their relatively favourable "vintage structure" of capital (because of investing in the riper fruits of the "creative diffusion process" both concerning products and related process technologies), non-metropolitan areas will be in a relatively favourable position and capture (part of) the market of producers located in metropolitan areas. Consequently, during this phase the indigenous dynamics and innovative activities (though now directed towards different types of innovations; cf. Fig.1) of the present technology system will shift to non-metropolitan areas[5].

Our conceptual-theoretical framework can be summarized in Fig. 2 which relates the evolution of spatial structures to the effects of "new technology systems" and "creative diffusion".

In Fig. 2 "relative" performance implies the "score" of a certain region on a certain sector (for example, share of employment in this sector) as compared with some average or "normal" level (for example, share in total employment). For the sake of simplicity we have sketched the ideal-typical performance of a metropolitan area vis-à-vis a peripheral area for sectors linked to technology systems of different periods.

Our previous framework would suggest a spatial classification of sectors according to phases of the related technology and "creative diffusion" cycle: In this respect we would expect metropolitan areas to perform better (with regard to their share in employment, production, innovations, etc.) for sectors linked to the most recent technology system, while peripheral areas are assumed to perform relatively better for sectors related to "former" technology systems. Our interpretation is clearly in contrast with the usual typology of metropolitan areas as general centres of innovation. In our framework, both

Figure 2: Spatial structures and new technology systems

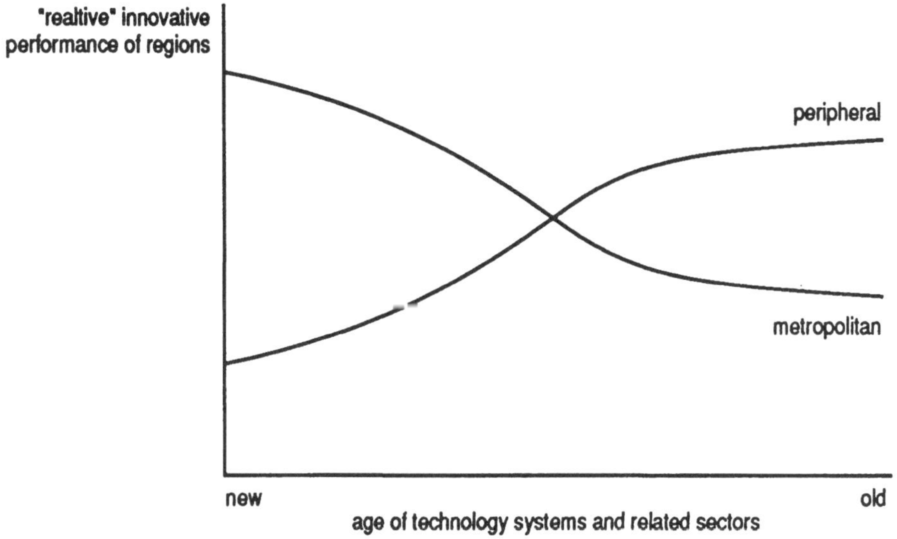

metropolitan and non-metropolitan areas are expected to perform their own specific innovative role during the "life cycle" of sectors and technologies.

We have presented our framework of analysis at an intranational level. Clearly, the international level could be introduced by suggesting a fourth phase (cf. Vernon 1966) during which the indigenous dynamics and innovative activities of the "technology system" shifts (from the non-metropolitan areas) to foreign countries. In the following analysis we will ignore such a phase, however.

2.4 Testing for Empirical Validity

Having outlined our "dynamic incubation" framework, we will now attempt to test the empirical validity (of some elements) of this framework. In this respect, it is noteworthy that an earlier analysis of the manufacturing sector in The Netherlands (Davelaar and Nijkamp 1987a,b,c) has demonstrated that for this sector the inherent dynamics, in terms of innovation potential (i.e the capacity to generate innovations), is nowadays located in non-metropolitan (i.e. intermediate and peripheral) zones. As the manufacturing sector as a whole is a rather "old" sector (i.e. the greater proportion of the firms being related to former technology systems), it appeared to be rather difficult to trace the exact space-time trajectory of specific manufacturing sectors[6]. Consequently, testing the empirical validity of especially phases 1 and 2 of our theory appeared to be far from easy. Besides, the determination of the space-time trajectory of firms related to more recent technology systems may be more interesting from the viewpoint of current growth perspectives. Given these considerations, in the following analysis we will mainly focus attention on the space-time trajectory (especially aiming at testing the validity of the first two phases of our framework) of selected producer service subgroups which are generally acknowledged to be linked to, and of utmost importance to, more recent technology systems.

3 Producer Services in The Netherlands

The aim of this section is to provide some empirical evidence for the relative "economic performance" of the producer service sector in The Netherlands compared with other sectors in recent years. As a first approximation, we will confine our analysis to those firms which belong to SIC-sector 84 (notably computer services, marketing offices, management consultancy etc.). Thus, we will exclude the banking and insurance sectors which also supply several intermediate producer services. The reason for doing so is that we consider these sectors to be relatively "old" sectors (i.e. compared with the SIC-sector 84) that are not dominated by "entrepreneurial innovation" and a resulting high level of new firm creation.[7] As it is our aim (in Sections 6 and 7) to analyze the space-time trajectory of innovative and dynamic sectors (in the sense of new "Schumpeterian" entries), we have excluded these latter sectors a priori from our analysis.

Clearly, any attempt to give an in-depth statistical analysis of the historical growth pattern of the producer service sector is very difficult. Data for this sector are relatively scarce compared with "old" sectors like agriculture and manufacturing (cf. Lambooy and Tates 1983). Also, statistical revisions and redefinitions hamper a profound time-series analysis of this sector regarding such variables as employment, value added and production. These problems become even more troublesome, if, as in our case, we want to study the developments within several subcomponents of the producer service sector. As a consequence of these statistical limitations our empirical analysis has several limitations.

The growing economic importance of the producer service sector is best illustrated by means of employment data which we derived from the Statistics of Employed Persons (i.e. excluding self-employed persons). These statistics are based on a sample, which is repeated every year by The Netherlands Central Bureau of Statistics. As can be seen in

Table 1 these data can, for various years, be obtained at a three-digit level. Our three-digit figures relate mainly to the 1970s as after 1979 employment data at a three-digit level have only been collected for firms employing more than ten employees.

It should be noted that before 1984 only jobs of 15 hours or more a week have been counted in these statistics, while from 1984 onwards also jobs with less than 15 hours a week have been included[8]. This explains the rather drastic increase in the total number of jobs between 1983 and 1985 depicted in Table 1. As a result of the increased importance of part-time jobs the direct link, however, between jobs and employment has been more or less lost, however.

The growth of employment (jobs) in the period 1973-1985 has indeed been rather uniform among the various (three-digit) subgroups of the producer service sector. Especially the growth of the computer services sector has been explosive. On the other hand, growth of employment in secretarial agencies has been of a fluctuating nature. In 1979

Table 1: Jobs in the producer service sector in The Netherlands

YEAR SECTOR	1973	1975	1977	1979	1981	1983	1985
Solicitor's offices (841)	7,876	9,089	10,390	10,831			12,685
Accountancy/tax advice (842)	29,375	30,250	37,109	45,880			63,013
Computer services (843)	61	1,217	2,824	4,723			13,615
Engineering (844)	33,851	35,053	41,912	48,440			51,637
Advertising (845)	6,565	7,295	6,640	8,782			10,197
Consultancy (846)	7,096	7,531	7,981	9,030			9,964
Press agencies (847)	569	588	412	598			661
Secretarial agencies (848)	27,852	37,671	38,732	47,086	38,500		102,430
Remaining categories (849)	22,611	21,347	20,222	23,556			28,219
Total number of jobs	3,841,241	3,914,055	3,966,337	4,117,403	4,161,108	3,976,735	4,480,510
Share of jobs in SIC-84 (%)	3.54	3.83	4.19	4.83	4.96	5.18	6.53

Source: Statistics of Employed Persons (several years)

this sector included 47,100 jobs, 53,600 in 1980, and only 38,500 in 1981. These fluctuations might partly be a reflection of the more general economic tide on which this sector seems to be dependent.

Also the share of jobs in the producer service sector as compared with the total number of jobs has been steadily increasing (see the last row of Table 1). Thus, Table 1 quite clearly demonstrates the growing importance of the various producer service subsectors as a pool of new jobs. It is noteworthy that the number of jobs in the manufacturing sector as a whole has been steadily decreasing during this same period. Figure 3 summarizes the development of jobs in manufacturing compared with the producer service subgroup as a "whole" (i.e. SIC-84).

Fig. 3 clearly depicts two opposite patterns. On the one hand, the steady rise of jobs in the producer service sector is clearly reflected. On the other hand, the manufacturing sector is rapidly losing its role as a jobprovider. Even the statistical redefinition of 1984 (i.e. also including jobs of less than 15 hours a week) did not alter the rather poor "performance" of the manufacturing sector during the period 1973-1985.

The increased importance of the producer service sector in total intermediate consumption of firms and in total net value added of firms is reflected in Tables 2 and 3 respec-

Fig. 3: Jobs in the producer services and manufacturing sector

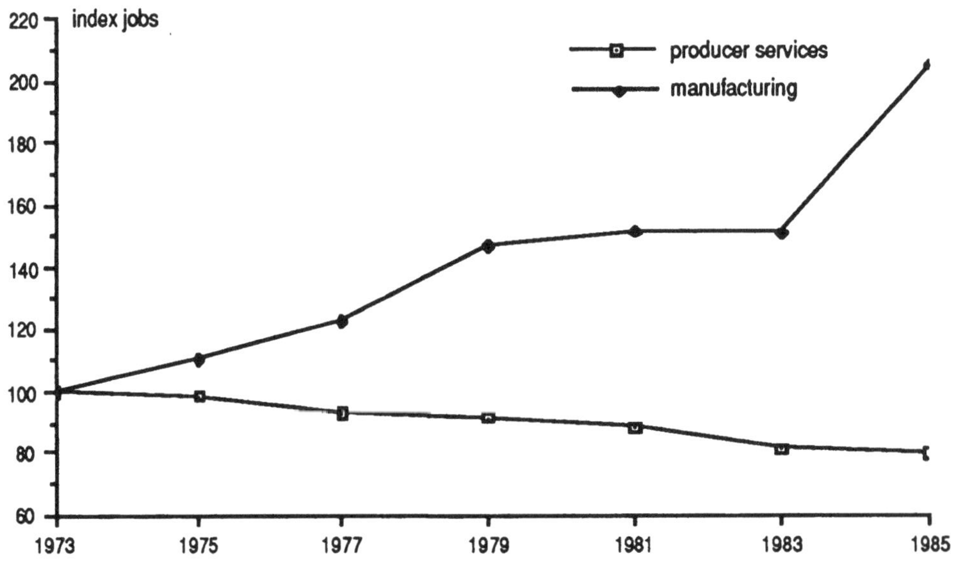

tively. In each table time-series have been constructed that are more or less comparable (denoted as "comparable" years)[9]. In interpreting these tables, however, one should be aware of the fact that the statistical "ignorance" effect (as mentioned before) of the producer service sector is also reflected in these tables, as the statistics are not available at a two-digit (SIC 84) level. Consequently, the data relate to an aggregation of SIC-84 and SIC-85 (letting offices)[10].

Both Tables 2 and 3 show that the general pattern is towards a growing economic importance of the producer service sector, in terms of the share in both intermediate deliveries and in total net value added. The increasing performance of this sector seems to have slightly stagnated in the years 1981 and 1982, but, as Table 1 reveals, this might to a large extent be due to the rather drastic decline of SIC-848 (secretarial agencies) in these years. However, it is not possible to fully test this assumption given the aggregation level of the available data. In 1983, however, both tables again show a "recovery" of the producer service sector. Thus, the hypothesis that producer services "perform an increasingly pivotal role as intermediaries in the production of both goods and services"(Daniels 1987:1) is indeed confirmed by our data.

The growth of producer service firms is often (partly) ascribed to a "crowding out" or "externalization effect". This implies that firms which previously performed such ser-

Table 2: Intermediate deliveries provided by the producer service sector

"Comparable" years	1972	1973	1976	1977			
Total intermediate consumption of firms (mlns)	78,390	90,390	13,0300	13,8780			
Amount provided by SIC 84/85 (mlns)	1,440	2,170	3,510	3,730			
Share provided by SIC 84/85	1.83	2.40	2.69	2.69			
"Comparable" years	1977	1978	1979	1980	1981	1982	1983
Total intermediate consumption of firms (mlns)	133,000	140,000	1530,00	167,000	178,000	198,000	205,000
Amount provided by SIC-84/85 (mlns)	6,480	7,490	8,330	9,290	9,810	10,378	11,067
Share provided by SIC-84/85	4.87	5.35	5.44	5.56	5.51	5.24	5.40

Source: National Accounts (several years)

Table 3: Net value added (market prices) of SIC 84/85

Year	Net value added SIC 84/85 (mlns)	Total net value added of firms (mlns)	Share SIC 84/85 in total net value added
1971	4,145	108,780	3.81
1972	4,582	122,610	3.74
1973	5,336	140,230	3.81
1974	6,340	157,640	4.02
1975	7,345	170,920	4.30
1976	8,830	196,640	4.49
1977	10,501	214,480	4.90
1978	12,170	230,900	5.27
1979	13,280	244,390	5.43
1980	14,250	259,800	5.48
1981	14,600	278,850	5.37
1982	14,948	283,900	5.27
1983	16,112	294,480	5.47

Source: National Accounts 1969-1981 (revisions for the years 1969-1976). National Accounts 1985 depreciation figures for the SIC categories have been derived from the original National Accounts 1973-1985

vices in-house now board out these services, because of either cost or expertise motives. In the first case costs would be higher should the firms perform these services in-house[11]. In the second case, however, the lack of expertise with respect to the tasks needed is the main driving force for externalization[12]. "Structural adjustments" due to the growing importance of information in the production process (cf. Naisbitt 1984; Noyelle and Stanback 1984), as well as the related innovations in the field of communication and information technology, have been a major driving force for the latter type of externalization effect. In this respect, de Haan and Tordoir (1986) have shown that especially in "high tech" sectors which exhibit high values added (i.e. sectors related to the more recent "technology system" like electronics, instruments and producer services themselves) the role of information in the production process has increased considerably in recent years. This is reflected in both a growing share of employees within these sectors being engaged in knowledge intensive functions and in a growing input of external producer services. Their results seem to indicate a "structural adjustment process" in which information (handling) functions become more important, both within firms and between them.

In the next section we will select special subclasses of the producer service sector which may be considered to be highly dynamic and largely affected by the growing importance of (the handling of) information in society (i.e. those producer service sectors which are linked to the recent information technology system, cf. Freeman 1987).

4 Selected Producer Service Groups

In the foregoing section we noticed already the scarcity of data on the growth of the producer service sector reflected inter alia in number of jobs, net value added and intermediate deliveries. At the spatial level however, these data limitations become even more severe. Besides, even at the two-digit level (for which some data are available) the producer service sector consists of rather heterogeneous subclasses. Consequently, the space-time trajectory of the (innovative subgroups of the) producer service sector is not easy to trace due to a serious lack of available data. Therefore, we had to construct our own data set for The Netherlands with the help of the Chambers of Commerce in order to carry out a more disaggregate, spatial and dynamic analysis. For this purpose we first selected special subclasses of the producer service sector that are linked to the more recent information technology system. The framework for selecting these subgroups is sketched in Fig. 4.

Thus, we have used two (interrelated) criteria for selecting and distinguishing two broad categories of producer service firms:

1. The services provided by these firms are knowledge-intensive, i.e. those producer service firms which are expected to provide external services for reasons of expertise.
2. The services provided are linked to the more recent "information technology" system. In this respect we made a distinction between services that cater to the needs of information itself (the "software component") and those services with the capacity to store, process and disseminate information within the firm (the "hardware component").

Recent developments with respect to CAD/CAM, CNC machines, information networks and so on, have largely influenced the services provided by the engineering and computer-services sector. As these services are to a high degree oriented towards increasing the mere information handling capacity (in firms and government institutions, for example); we have labelled these services the "hardware" sector.

Figure 4: Technological change and producer services

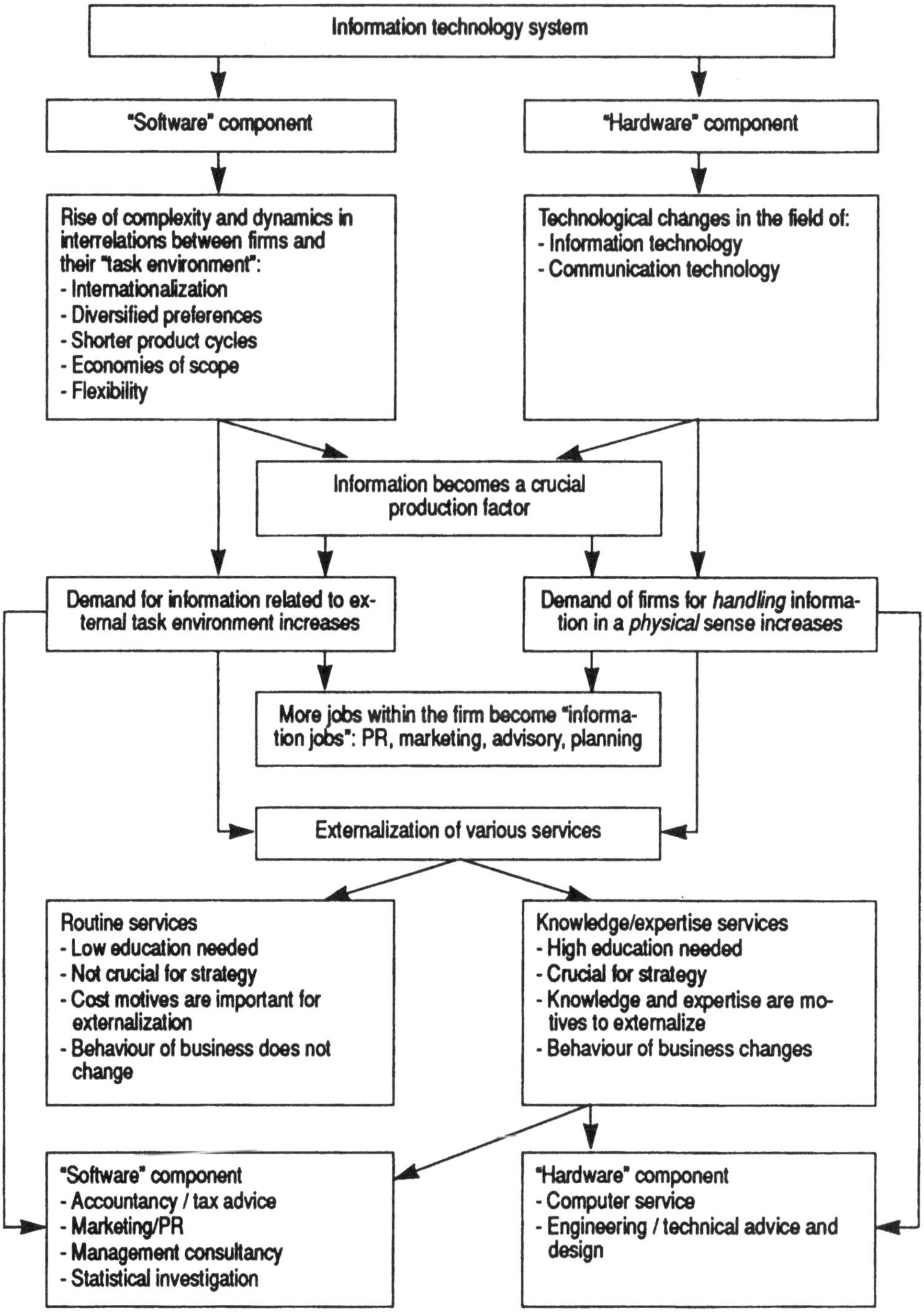

The generally increasing need for information because of dynamic and complex changes in firm-"task" environment relations (cf. Lambooy and Tordoir 1985) has been an important impulse for producer services acting on the edge of this firm-"task" environment. As these firms are not principally concerned with increasing the information handling capacity of firms they are labelled here the "software sector". This has also been the reason for including "management consultancy" in this category, although part of its more recent economic performance might also be related to the effects of changes in the "hardware component" within the firm (i.e. changes needed in organization structures as a result of the implementation of new information systems within the firm (cf. Huppes 1987).

In the next section the characteristics of our empirical data base will be discussed in more detail.

5 The Empirical Data Base

On the basis of the criteria described in the foregoing section, we selected the following provisional sectors at the four-digit level:

"Software" subgroup:

	SIC-code:	Sector:
Financial information:		
	8421	Accountancy
	8423	Tax advice
Non-financial ("general") information:		
	8451	Advertising
	8461	Market analysis
	8462	PR advic
	8464	Statistical analysis
	8466	Management consultancy

For reasons to be indicated later on, we have made a distinction between firms mainly performing financial and non-financial services.

"Hardware" subgroup:

	SIC-code:	Sector:
Computer services:		
	8431	Computer services

Engineering sector:

8441	Engineering/technical advice and design
8443	Construction advice and design
8449	Remaining engineering advice/design

The data situation for the producer service subgroups (especially at the three-digit and four-digit levels) is rather poor in The Netherlands, especially if one wants to study space-time patterns. Because of confidentiality the Central Bureau of Statistics does not publish such data on a very disaggregate (regional) level. Consequently, in order to circumvent these data problems we had to construct our own data base. In this respect, the data collected by the Chambers of Commerce and centrally available at the Central Data Bank in Woerden, appear to be the best substitute[13]. In The Netherlands, (nearly) every establishment, whether head or subsidiary establishment, has the legal obligation to register itself with the relevant Chamber of Commerce.

Thus, these Chambers not only possess data on number of establishments, but also on number of employees, main activities of the establishment (SIC codes), location and so on[14]. The Central Data Bank not only collects data from the regional Chambers, but also employs other sources (e.g. inquiries) in updating its data base.

We, therefore, decided to use this data base with respect to our selected producer service subgroups. In order to minimize costs, we constructed the following sample design with respect to size categories and sectors (see Table 4). Figures without parentheses refer to the situation as at *May 1, 1987*, when the Central Data Bank made several cross-section counts. Figures in parentheses are based on our own *sample* results which represent the (December) 1987 situation. Data were gathered at the *establishment level*, while only small establishments (i.e. less than five employees) have *not* been integrally observed in our data base. It is evident that the number of establishments (and the related employment in these sectors) appears to have increased considerably even in the relatively short period May-December 1987! This applies especially to the computer services and "general information" group. In these cases the number of establishments with more than four employees (which were integrally observed in our sample) increased considerably in this period. Also the observed number of establishments in the size categories one and two to four employees (for which we developed a stratified sample design, see Table 4 above) clearly exceeds the number that might be expected on the basis of the (total) May figures presented by the Central Data Bank. As a matter of fact, considering these latter figures as the total population of a repeated Bernoulli experiment, a sample outcome as we ob-

Table 4: Sample design for selected producer service subgroups

Number of employees	Probability of being selected	Total number of establish-ments	Financial information Group 1	Computer service Group 2	Engineering advice/design Group 3	General information Group 4
1	0.1	11,704	1,019 (111)	2,352 (263)	2,805 (297)	5,528 (625)
2-4	0.333	4,179	365 (125)	795 (290)	995 (328)	2,024 (713)
>4	1.0	3,072	427 (459)	666 (751)	1,055 (1,136)	924 (1,004)

served for these two subgroups would be highly implausible (affirming that the increase in the number of establishments in these size categories is statistically significant)[15].

In the accountancy and tax advice and engineering advice sector on the other hand, development appears to be less expansive though still growing. In this case the number of establishments with more than four employees have increased in both sectors. Also the number of establishments with one employee seem to have increased in both sectors in this period (though not in a statistically significant way). In the size category two to four employees, however, the observed sample outcome corresponds to the "expected" outcome on the basis of the May figures. Consequently, our hypothesis that the selected subgroups can be considered as highly dynamic (i.e. many entries) and rapidly increasing in economic importance (i.e. belonging to the new information technology system), is indeed confirmed by the data. These signs of highly dynamic sectors are also reflected in Table 5 in which the estimated number and percentage of "new" establishments, i.e. those establishments with a date of establishment after January 1 1985, is presented[16].

In the first row of Table 5 the estimated total number of establishments is presented. In the second row the estimated total number of "new" establishments (i.e. with a date of establishment after December 31, 1984) is given[17]. Next, the third column contains the total estimated number of establishments with more than one employee, while row 4 contains the estimated number of "new" establishments within this latter size category[18]. Next row 5 contains the (estimated) percentage of "new" establishments, while row 6 contains the same percentage concerning establishments with more than one employee.

From Table 5 it is again confirmed that computer services, "general information" and "engineering" are (and in this order) highly dynamic sectors compared with accountancy (and architects)[19], for example. Those sectors are growing rather rapidly and are characterized by high entrance rates as a result of entrepreneurial activity (innovation). In this

Table 5: "New" establishments in selected producer service sectors

	Financial information	Computer service	Engineering advice/design	General information
Estimated total number of establishments Dec. 1987	1,944	4,251	5,090	9,393
Estimated total number of establishments > 1 employee	834	1,621	2,120	3,142
Estimated total number of "new" establishments	489	2,300	1,710	3,983
Estimated total number of "new" establishments > 1 employee	79	710	460	913
Percentage of "new" establishments (row 3 / row 1)	0.25	0.54	0.34	0.42
Percentage of "new large" establishments (row 4 / row 2)	0.09	0.44	0.22	0.29

respect, these sectors could indeed be considered as forming part of a "new technological system" (clustered around telecommunication and information technology).

Even with respect to establishments with more than one employee, the last row in Table 5 reflects the highly dynamic and successful nature of these three subgroups. In this latter case the share of establishments younger than three years amounts to 44%, 29% and 22% in the computer services, "general information" and engineering sectors respectively. The figures with respect to "financial information" (and architects), on the other hand, exhibit a far more static pattern.

Thus, we can conclude that the computer services, "general information" and engineering sectors (in this order) can indeed be considered as dynamic and relatively young sectors performing new tasks which are becoming increasingly important in the economic system. In this respect these sectors could fall under the heading "entrepreneurial innovation" (cf. Bearse, 1978). In subsequent sections we will concentrate on the space (-time) trajectory of these latter three, highly dynamic (and innovative) subgroups.

6 Spatial Dispersion of Selected Producer Service Sectors

In the present section we will especially concentrate on the spatial dispersion of the computer services, "general information" and engineering sector. The labels "new" (i.e. many entries) and "innovative" appear to be best suited for these subgroups. Clearly, a certain fraction of these establishments, especially in the engineering and "general information sector[20], might still be rather old. At the sectoral level the notions "new" and "innovative" are relative concepts (i.e. compared with other sectors).

Davelaar and Nijkamp (1987a,b) also analysed the regional pattern of "innovativeness" and "innovation potential" of the manufacturing sector. It was found that concerning the *innovation potential* of manufacturing establishments, the western (metropolitan) part of The Netherlands appeared to be lagging behind nowadays. Looking more specifically at "innovativeness" (output of the innovation process), however, the performance of the western part appeared to be far less devastating concerning high quality *product* rather than *process* innovations (Davelaar and Nijkamp 1987c).

As the geographical scale for the producer service subgroups, we used the so-called COROP level as the spatial scale of our analysis. The COROP level is based on a standard statistical demarcation of The Netherlands into 40 nodal regions. For the sake of convenience, but also for statistical reasons (in some COROPs the number of observations in the sample may be quite low), we have amalgamated these 40 COROPs into the following areas[21]:

1. "Metropolitan" COROPs: Greater Amsterdam Area, The Greater Rotterdam Area, The Metropolitan Area of The Hague, Utrecht and Arnhem-Nijmegen
2. "Spillover" COROPS surrounding the metropolitan COROPS: These include Veluwe, Achterhoek[22], Zuidwest-Gelderland, Kop van Noord-Holland, Alkmaar, Ijmond, Haarlem, Zaanstreek, Gooi en Vecht, Leiden, Delft en Westland, Oost-Zuid-Holland, Zuidoost-Zuid-Holland and ZIJP
3. COROPS located in the province of Brabant
4. The southern peripheral areas COROPS located in the provinces of Zeeland and Zuid-Limburg.
5. The northern peripheral areas: COROPS located in the northern provinces of The Netherlands: Groningen, Friesland, Drenthe and Overijssel.

Our empirical results are based on "estimated" employment data concerning the three producer service subclasses distinguished. These estimations are based on our sample derived from the Chambers of Commerce described in the foregoing section. Consequently, the estimations represent the situation at the end of 1987. In making these estimates we multiplied the observed employment figures from the sample by a factor 10 and 3 respectively for establishments with one and two to four employees (being equal to the inverse of the chances of being selected).

As the Central Data Bank is not allowed to report the exact number of employees within the establishments, the following (employment) size categories are used: 1, 2-4, 5-9, 10-19, 20-49, 50-99, 100-199, 200-499, 500-749, 750-999, and >999. Therefore, we used the median of these categories in estimating our employment figures. As none of the establishments in the selected subclasses exceeded the "critical" threshold of 1000 employees, the size category of 1000 and larger presented no problem in this respect.

Table 6 shows the patterns of (estimated) regional employment in our three subgroups distinguished.

It is noteworthy that the ordering of the regions in Table 6 is based on a distance decay pattern from centre to periphery. In column 1 the regional share of (total) commercial service jobs (in the SIC categories 6,7,8) has been given, while in the second column the same has been done for the manufacturing sector[23]. A first conclusion that can be derived from Table 6 is the metropolitan orientation of the commercial service sector as a whole compared with manufacturing. This "metropolitan bias", however, applies even more so for our selected "new" and "innovative" subsectors of the producer service sector! The five metropolitan COROPS are estimated to account for more than 50% of employment in each of the three producer-service groups distinguished!

On the other hand, both the northern and southern peripheral areas exhibit the opposite pattern. They attract a considerably smaller fraction of total commercial service employment compared with their manufacturing share. However, their "performance" concerning the producer service sectors distinguished is even more negative. While attracting (together) about 30% of manufacturing employment, their share of employment in the computer services and "general information" sector is estimated to be only 10%! The only exception to this appears to be the "score" of the northern peripheral area with respect to the engineering sector (according to Table 5 this sector was the most "static" of our selected subgroups). The "spillover" region appears to be the most "homogeneous" area in terms of employment shares of the various sectors distinguished in Table 6: all

Table 6: Regional employment within producer service sectors

Region	Regional share of employment in services (SIC 6,7,8)*	Idem manufacturing*	Regional share of estimated employment in computer services Dec. 1987	Idem "general information"	idem "engineering"	% of employment within region in branch plants	% of population in 1983**
Metropolitan	44.8	24.7	54.9	54.8	51.1	10.0	31.3
"Spillover"	22.3	24.4	23.1	20.6	19.8	7.5	25.3
Brabant	12.2	20.8	11.7	13.3	9.9	18.2	14.6
Southern Peripheral	7.7	12.1	3.1	3.5	5.0	17.2	10.0
Northern Peripheral	13.0	17.9	5.8	7.7	14.0	23.5	18.3
(Estimated) total employment	1,538,000	941,700	28,749	31,571	42,452		

Source: * Statistics of Employed Persons 1986
** Regionaal Statistisch Zakboek,1984: 229

(estimated) shares could be located in an interval between approximately 20% and 24%. The Brabant region, finally, appears to attract a "normal" (i.e. compared with its share in total commercial service employment) fraction of employment in our selected producer service sectors. However, this fraction is considerably less than its manufacturing share.

In column 6 of Table 6, we have tried to gain some further insights into the question whether a regional bias concerning the "branch plant effect" can be identified. For this purpose we have estimated the regional shares of employment in "branch plants". As only about 10% of the establishments in the sample could be designated as "branch plants", we estimated an aggregate (i.e. the summation of the three selected sectors) figure in this column. From column 6 of Table 6 it can be concluded that the branch plant effect indeed exhibits a regional bias. This effect shows a center-periphery pattern. In peripheral areas the share of employment in branch plants is clearly larger than in metropolitan areas. In this regard, the Brabant, southern and, especially, the northern peripheral regions "perform" much better concerning their share of "branch plant" employment. The "metropolitan" and "spillover" areas show, however, the opposite pattern. This implies that in the more peripheral areas the producer service sectors are not only underrepresented, but they also depend more upon external initiatives than is the case in the more central regions! On the other hand, however, it is clear that only a minority fraction of total producer service employment within these regions is of the "branch plant type" (stressing also the importance of an indigenous component in these areas).

Thus far we have only presented a rather static description of the spatial dispersion of producer service sectors. The time dimension may, however, be equally interesting. Consequently, in the next section the space-time dimension will be dealt with.

7 The Space-Time Trajectory of the Producer Service Sectors

In this section the spatial-dynamic aspects of the producer service subsectors will be analysed. There is a serious lack of data in The Netherlands on (comparable) time-series and spatial data with regard to the producer service sectors. For example, statistics from which one could derive the space-time trajectory of the computer services sector are completely lacking. Such an analysis would only, to some extent[24], be possible at the two-digit level. Unfortunately, the Chambers of Commerce are not able to provide relevant data for all years on the spatial dispersion of producer service establishments. This is caused by the fact that establishments which existed in 1970, but stopped business later on, were removed from the data set. Consequently, only a momentary picture can be presented by the Central Data Bank, and we will, therefore, have to adjust our own sample in order to derive some, tentative conclusions. For this purpose the variable "date of establishment" is crucial in our analysis, as this variable can be used to approximate the spatial evolution of the sectors distinguished. We constructed the following indicators (which will be calculated for each of the three subgroups separately):

1. Regional share of employment in establishments that have a "date of establishment" earlier than January 1, 1976
2. The same with respect to a "date of establishment" earlier than January 1, 1981
3. The same with respect to a "date of establishment" earlier than January 1, 1985
4. Regional share of employment in establishments that have a "date of establishment" later than December 31, 1984

Indicator 1 might, under certain conditions, be a reflection of the situation prevailing at January 1, 1976, and indicators 2 and 3 with respect to January 1, 1981 and January 1, 1985, respectively[25]. To be sure that these indicators are an adequate reflection of the spatial dispersion of employment prevailing at these dates, the following conditions have (inter alia) to be satisfied:

1. Regional (total) growth of employment within establishments that already existed in 1976, and still existed in December 1987, should be the same in the period 1976-1987.
2. Regional "death rates" for (total) employment within establishments, that existed at 1, January, 1976, but stopped business later on, should be the same over this period.
3. Interregional relocation effects should cancel each other out during this period.
4. Regional "shifting" of sectoral employment patterns due to establishments changing their main activity (i.e. changing their SIC-code) should be equal.

Of course it is difficult to determine the validity of these conditions, but it is clear that these conditions will become more problematic the longer the time interval considered and the smaller the region analysed. Some remarks, however, seem to be appropriate in this respect. First, since we have amalgamated the 40 COROPs into five broad groups, these conditions are less restrictive than in the case of individual COROPs. Secondly, it appears that conditions 2 and 3 are especially biased against the "performance" of metropolitan areas. Concerning condition 2, for example, Wever (1984) has demonstrated that the "death rates" of new establishments appear to be the highest in metropolitan areas. Regarding condition 3, the incubation hypothesis in its complex version (cf. Leone and Struyck 1976), as well as empirical research in The Netherlands (Buunk and Elderman 1982; Molle and Vianen 1982) has demonstrated that in the metropolitan areas a net out-migration of employment can be identified. The total effect of these relocations on regional employment patterns, however, appears to be rather small (cf. Potters 1983; Bearse 1978). Thus, these studies might suggest an underestimation of the "performance" of metropolitan areas on the basis of our above-mentioned indicators (at the prevailing dates). Concerning the "direction" of the (possible) regional "bias" of the indicators 1 and 4 no a priori judgements can be made.

Indicator 4 is of course intended to reflect the more recent spatial development patterns of our producer service subgroups. By means of this indicator it can be analysed whether our selected subsectors are improving or worsening their relative "performance" in the past three years.

The results of our analysis are extremely interesting and point at the same type of space-time trajectory of these "new" and "innovative" producer service subgroups. Consequently, in this section we will discuss the patterns for the computer services sector in detail, while only some general remarks will be made concerning the "general information" and "engineering" (see Tables 7 - 12).

Table 7: Space-time trajectory of the computer service sector

	Approximated share of employment in pre–1976 establ.	idem with respect to pre–1981 establishm.	idem with respect to pre–1985 establishm.	Approximated share of "new" employment	Approximated share of total employment Dec. 1987	Regional share commercial service employment SIC 6,7,8*
Metropolitan	64.5	62.3	56.2	50.8	54.9	44.8
"Spillover"	16.3	18.4	23.0	23.5	23.1	22.3
Brabant	9.2	11.0	11.6	13.2	11	12.2
Southern Peripheral	1.5	1.9	2.5	5.1	3.1	7.7
Northern Peripheral	6.1	5.0	5.5	7.0	5.8	13.0

Source: * Statistics of Employed Persons 1986 (Tables relating to these subsectors can be found in the Appendix). Approximations do not sum to 100 because of rounding errors and establishments with "unknown" location

If we accept indicators 1-4 as (rough) approximations of the spatial dispersion of employment prevailing in 1976, 1981, 1985 and 1985-1987, respectively, some interesting patterns can be found. In the first place, the computer services sector (still) appears to be highly concentrated in metropolitan areas, a result also found by Koerhuis and Cnossen (1982). As time passes and the "new technology system" becomes more and more integrated into the economic structure, however, this dominant position is gradually eroded, as suggested by the indicators used in Table 7. However, the share of employment in establishments that have been set up in the period 1 January, 1985 to December 1987 (and still existing in December 1987), continues to exceed the average "normal" level[26] in the metropolitan areas. But, in this case the score on indicator 4 is considerably below the scores on the other, historically oriented, indicators. This pattern of initial strong concentration in metropolitan areas versus (slow) spatial deconcentration is fairly uniform among individual metropolitan COROPS (see Table 8)[27].

Table 8: Computer services in metropolitan areas

	Approximated share of employment in pre–1976 establ.	idem with respect to pre–1981 establishments	idem with respect to pre–1985 establishments	Approximated share of "new" employment	Regional share commercial service employment SIC 6,7,8*
Arnhem / Nijmegen	8.4	5.7	5.1	2.8	4.1
Utrecht	19.5	17.8	18.5	16.0	7.9
Amsterdam	20.3	20.0	16.7	16.3	14.0
The Hague	11.4	8.1	7.0	6.7	6.8
Rijnmond	4.9	10.7	8.9	9.0	12.1

Source: *Statistics of Employed Persons 1986

Our indicators suggest that the Rijnmond area is clearly lagging behind compared with the other metropolitan regions distinguished. In the Rijnmond case, none of our indicators exceeds the "normal" level, taken to be the "overall share" in commercial service employment. Also concerning the share of "new" employment (indicator 4) the Rijnmond area still performs rather poorly. In comparing the indicators 1 to 4, it is plausible to assume that the other metropolitan regions all started from a high level (compared with their "normal" level), while slowly approaching their "normal" level (indicator 4). However, the Utrecht area still appears to attract a share that considerably exceeds its "normal" level[28]. Also the Greater Amsterdam area still succeeds in attracting a higher than normal share of new employment.

The second region in our analysis, which seems to have passed its normal level, appears to be the "spillover" region which by 1985 (as reflected by indicator 3) has apparently passed its "normal" share of computer services employment.

The third region in order, appears to be Brabant which attracted an above-normal share of new employment (indicator 4). On the other hand, the southern and, especially, the northern peripheral areas have not yet succeeded in attracting a normal share. Also their shares of new employment are considerably below their shares in total commercial service employment. In comparing the performance of these regions on the indicators 1 to 4, however, the southern peripheral area appears to be more rapidly expanding its share than the northern peripheral area.

Consequently, the space-time trajectory of the computer services sector, as reflected by our indicators, appears to be clearly developing along the lines of our theoretical framework for sectors linked to "new technology systems" sketched in section 2. In comparing the indicators 1 to 4, it is clear that at first a very strong metropolitan concentration can be observed (phase 1), while nowadays the pattern seems to be shifting towards non-metropolitan areas (phase 2).

Concerning the engineering sector (see Tables 9 and 10 in the Appendix), the following conclusions can be drawn:

The engineering sector is the most static sector of our selected subgroups, i.e. in terms of the importance of new firm formation rates (cf. Table 5). Consequently, we expect the space-time trajectory of this sector, along the lines of our "dynamic incubation" framework, to be further developed than the other sectors selected.

This is indeed confirmed by the data as regional shares appear to be less fluctuating (compared with computer services and general information), while these shares are also more closely related to the normal shares of commercial service employment (as can be seen by comparing columns 5 and 6 of Tables 9 and 10).

Metropolitan areas (again) appear to be continuously losing their dominant position, while especially the "spillover" and Brabant regions appear to be improving their relative performance (also, the pattern within the metropolitan areas is rather uniform as can be seen from Table 10).

All metropolitan COROPs distinguished attract a smaller proportion of new employment compared with their share of pre-1985 employment. In case we would exclude COROP region Delft from our analysis (with respect to indicator 4), this conclusion would not hold for the COROPs The Hague and (again) Utrecht. Concerning the shares of new employment, the Amsterdam, Rotterdam (Rijnmond) and Arnhem/Nijmegen areas are nowadays already performing below their normal levels.

While the southern peripheral area is gradually improving its performance on the basis of our indicators, developments in the Brabant and "spillover" regions are much more explosive. Concerning their share of new employment, both the Brabant and "spillover" areas[29] attract a proportion which is larger than their shares in total commercial service employment. The southern peripheral region has not (yet) attained this normal level. Although the northern peripheral region scores reasonably well on the historical indicators (i.e. 1 to 3), the score on the new employment indicator 4 is rather poor.

Our indicators suggest that concerning the engineering sector the metropolitan areas as a whole experience an erosion of their dominant position. The opposite pattern holds for the "spillover" and Brabant regions. The southern peripheral region is improving its performance, although it has not yet succeeded in attracting a normal level. These conclusions are exactly in line with the theoretical model which we discussed before. The only exception to this appears to be the northern peripheral area, which already performed above normal with respect to the historical indicators 1, 2 and 3. The score of this region on indicator 4, however, points to a rather sharp deterioration of this favourable position[30].

Last but not least, concerning the space-time trajectory of the general information sector, the following conclusions can be drawn (see Tables 11 and 12 in the Appendix):

As before, the general information sector appears to be heavily concentrated in metropolitan areas. Again, however, this dominant position is being eroded.

Looking more specifically at the metropolitan areas, Table 12 shows that especially the Amsterdam region appears to have been the "incubation area" of this sector. Nearly 40% of employment in "old" establishments (i.e. which have a date of establishment earlier than 1 January, 1976) are located in the COROP Amsterdam! The results for indicators 2 to 4, however, point at a deterioration of this dominant position. As a matter of fact, this region nowadays attracts only a normal share of new employment, as reflected by indicator 4.

Also remarkable (again) is the rather poor performance of the Rijnmond (Rotterdam) area which does not exhibit a relatively high score on any of our indicators, while its share of employment in new establishments is also considerably below normal.

The other metropolitan areas, on the other hand, still appear to be in a kind of boom phase. In this respect the (total) shares of the metropolitan COROPs Arnhem/Nijmegen, The Hague and Utrecht in new employment (indicator 4) is 40% higher than their share in pre-1985 (indicator 3) employment!

Although expanding more smoothly, also the Brabant and "spillover" areas appear to have attracted a normal share of employment in this sector in the course of time.

The peripheral areas, on the other hand, although also expanding, again exhibit below normal scores on our indicators. The score of the northern peripheral area on indicator 4 can especially be ascribed to the province of Groningen (4.9% of total new employment in general information is estimated to be generated in this province).

8 Conclusions

In this paper the concepts of "new technology systems" proposed by Freeman et al. (1982), and the related concept of "innovation cycles", have been integrated into a "dynamic incubation theory". This theory may be useful in describing the space-time trajectories of new innovative sectors (and the related types of innovations; cf. Davelaar and Nijkamp 1989). According to this framework new "innovative" sectors may especially start from metropolitan areas onwards. Particularly in the beginning of a new technology system the generation of new "Schumpeterian" firms and related new products and services will be accentuated (cf. the innovation cycle concept). Because of both demand and supply conditions the metropolitan areas are then expected to be the "breeding place" of these new innovative activities.

By considering technological change as a continuous process, i.e. a "creative diffusion" process, this development will act as a driving force itself for the spatial shifting of indigenous dynamics to non-metropolitan areas. This applies especially to sectors which become more (though not completely) standardized in terms of the "products" (or services) generated. Consequently, in later phases of a technology system, part of the market possessed by "metropolitan producers" will be captured by "non-metropolitan producers". Continuous technological change may be one of the main driving forces in this spatial evolution. Consequently, in those phases especially the latter areas will perform relatively better (in terms of both economic variables and related innovative activities).

Previous research on the manufacturing sector in The Netherlands (cf. Davelaar and Nijkamp 1987a,b) has demonstrated that regional discrepancies in innovative activities may be an important driving force in the urban-rural shift of manufacturing activities. The same underlying (technological) forces may be at work in the internationally observed urban-rural shift of the manufacturing sector (cf. Keeble et al.1983; Keeble and Wever 1986). This shift cannot only be ascribed to the well-known branch plant effect (cf. Holland 1976), but also contains an important indigenous component as, for example, found in Bluestone and Harrison (1982).

Concerning the manufacturing sector statistical considerations hampered a rigorous test of the space-time trajectory of (subclasses) of this sector. Consequently, in this paper our main purpose has been to trace the space-time trajectory of sectors linked to more recent "technology systems", i.e. selected producer service subclasses. In this respect we have tried to determine whether spatial development of firms linked to the new "information" technology system prolongs along the lines of the urban-periphery trajectory as expected on the basis of (the first phases of) our framework. To this purpose first of all it has been demonstrated that this sector, and especially the three subgroups which we finally selected, is indeed characterized by a rapid expansion of value added, employment and number of establishments. For an identification of the space-time trajectory of these sectors, the lack of relevant data forced us to construct our own data set from the statistics provided by the Chambers of Commerce. In this respect we were able to calculate several proxy indicators which might serve as a reflection of the spatial dispersion of employment in these sectors at certain moments of time.

In this manner we tried to approximate the space-time trajectory of these new sectors. The results clearly support our "dynamic incubation theory". In initial phases, each of the producer subsectors distinguished appear to be heavily concentrated in metropolitan areas. In later phases, however, this dominant position becomes eroded and nowadays especially the intermediate zones (i.e. the "spillover" and Brabant areas) are improving their relative performance. Peripheral areas (i.e. both the northern and southern peripheral areas) still appear to attract a small proportion of employment in these sectors. Consequently, the space-time trajectory of our selected producer service subsectors clearly follows the line of the centre-periphery model depicted in the first two phases of our theory. As a matter of fact this trajectory appears to be most pronounced for the most static of the selected subsectors, i.e. the engineering sector. Concerning the most dynamic subsector (i.e. the computer services sector), employment still appeared to be most heavily concentrated in metropolitan areas. The score of the peripheral areas, in this case, however, appeared to be (extremely) poor in comparison with their scores for commercial service employment in general.

These results clearly confirm the previously sketched theoretical framework. However, it has also become clear that this framework should not be interpreted as a "static trajectory" that is repeated in the same way every time a new technology system enters the scene. Being a metropolitan area is no guarantee for attracting a large share of new and "innovative" firms in the first phases of the "emergence" of a new technology system. The Rijnmond area illustrates this latter statement quite clearly, as was shown by the rather poor performance on each of our estimated indicators. Consequently, there is no guarantee that all individual metropolitan areas will be "market leaders" concerning the next "technology systems". In this respect, the "creative diffusion" process itself will act as both a driving and selecting mechanism. This mechanism may be one of the most important, while largely ignored, driving forces of the competitive power and economic restructuring of spatial systems.

Footnotes

1 Consequently, in their opinion there is no unique relation between the phase of the economic "long wave" and the timing of this "swarming" effect.

2 A more recent example of this refers to the availability of fibre-glass networks which in the first instance will be developed in metropolitan areas (cf. Moss 1985).

3 This may be effectuated by means of the rise in branch plants, by relocations and by endogenous generation of new firms. Empirical research, however, points to a rather minor effect of the interregional relocation component (cf. Bluestone and Harrison 1982, Wever 1984). The importance of the branch plant component will, inter alia, be determined by the degree to which oligopolistic market structures will prevail during later phases of the related product life cycles. However, in these later phases at least part of the production will be realized in (small) independent firms.

4 On the basis of cost performance criteria.

5 Because of the embodied continuous "creative diffusion" process this framework reflects a more optimistic point of view for non-metropolitan (peripheral) areas than is usually suggested by the conventional filtering-down theory. Implicyty, the latter theory suggests that technological progress has come "to rest" before production shifts (because of non-technological factors such as labour costs, pollution regulations, unionization etc.) to non-metropolitan areas. This ignores the creative and innovative role (and the resultant competitive power) of non-metropolitan regions during later "life cycles" of technology systems. In our framework the "creative diffusion" process lies at the heart of the spatial shift of both (mutually reinforcing processes of) production and innovation (and not only of production as perceived in the filtering down theory).

6 In addition also the lack of disaggregate regional time series data is troublesome in this respect.

7 These sectors are evidently already characterized by high rates of market penetration.

8 As a consequece of the increased importance of part-time jobs.

9 This is caused by the fact that statistical revisions had a rather severe impact on the producer service statistics.

10 Fortunately SIC-85 is only small in term of employment (i.e. about 2 %) compared with SIC-84.

11 In this case, the firm could reach the same "quality level".

12 In this case, the "quality level" of the services performed inhouse would not be the same.

13 This applies especially to research performed at the level of individual establishments, as was also confirmed in a comparative study by SKIM (1981).

14 This does not imply, however that this data base can be considered to be error-free (cf. Wever 1984). During 1984 the Central Data Bank considerably improved the quality of its data base, for example, by removing non-existent establishments and improving SIC coding.

15 This can be understood by reconsidering the first two moments of the binomial distribution (as a matter of fact our sample can be considered as a repeated Bernoulli experiment).

If N = total population size and if we have a repeated Bernoulli experiment in which every member of the population has a chance p of being selected (and 1-p of not being selected) we have:

E(x) = expected outcome (i.e. number of successes) = p * n
V(x) = variance of outcome = p * (1-p) * n

Taking computer services as an example and considering the establishments with one employee, we have:

N = 2352 (i.e. the total May figure of the Data Bank) p = 0.1,
E(x) = 235, V(x) = 211

By constructing a 95% confidentiality interval around E(x) (i.e. 235 + 1.96 * 14.5), it can be seen that an outcome of 263 (as we observed in our sample) would be highly implausible given (a supposed) total population of 2,352. This holds even more so for the observed sample sizes within the general information sector (for both size categories) as well as the computer services sector (for establishments with 2-4 employees). Consequently, our conclusion must be that especially with respect to these sectors (Concerning all three size categories distinguished) the total number of establishments has increased during this relatively short period.

16 Giving the sample fraction, the number of establishments in the size categories 1 and 2-4 employees have been multiplied by 10 and 3 respectively.

17 A small fraction of this may be ascribed to establishments changing their location to a different Chamber of Commerce wich results in an new date of establishment. In general this effect will only be minor (cf. Wever 1984), especially as we are only considering a relatively short time period (less than three years).

18 This might be considered as a proxy for the degree of success of new establishments, i.e. they succeeded in passing a "critical" threshold of one employee (self-employed people).

19 For these subsectors similar estimations could be constructed which pointed at a pattern even "worse" than within the "financial information" group. Although

they are included in SIC-84, this is again a clear illustration of the heterogeneous nature of the subsectors included in this two-digit SIC category.

20 This may also apply to the computer services sector as "old" establishments may change their main activities in the course of time.

21 A"standard" classification of The Netherlands into central, intermediate and peripheral would not be appropriate in this case, as the large cities in the western part of The Netherlands show a rather distinct pattern compared with their immediate surroundings (as will become clear later on).

22 Although not really a "spillover" area of the Western Netherlands, this COROP has been included in this set because the observed patterns appeared to be rather consistent with this group and inclusion in remaining sets may be even more unrealistic. In any case, this choice is rather irrelevant for the empirical results that will follow.

23 It should be noted that these figures have been derived from the Statistic of Employed Persons 1986. As stated before, these statistics exclude self-employed persons. Consequently, the total figures may present an under-estimation. The "bias" concerning regional shares, however, will most probably be rather weak as self-employment will only be a minor fraction of total employment. Besides, there is no a priori reason for expecting the regional pattern of self-employment to differ significantly across regions.

24 That is, by using the Statistics of Employed Persons which does not count the number of self-employed persons.

25 These figures at least reflect regional shares in establishments of a certain age that have not relocated across Chamber of Commerce borders (as stated before, this may be a rather minor percentage).

26 In comparison with their share in total commercial service employment.

27 As said before, one should be careful in interpreting the data at such a disaggregate spatial level, as the indicators 1-4 may become less reliable in this respect.

28 Although not exactly comparable in terms of relevant periods, the results of Koerhuis and Cnossen concerning 1981 suggest that our estimates on the total "performance" of metropolitan areas in 1981 (indicator 2), as well as the individual results of the Amsterdam and Rijnmond areas, are very reasonable indeed. On the other hand, the performance of the Utrecht area with respect to indicator 2 may have been overestimated, while that of the Hague region may have been underestimated. This would imply that we are underestimating the "improved performance" of the first region, while underestimating the relative decline of the latter region in the recent decade.

29 In case we exclude the COROP region Delft this figure becomes a little smaller than the "normal" level in the "overspill" area.

30 This might point to a favourable position that is especially based on establishments linked to (or generated by) former technology systems.

Appendix

Table 9: Space-time trajectory of the engineering sector

	Approximated share of employment in pre–1976 establishm.	idem with respect to pre–1981 establishm.	idem with respect to pre–1985 establishm.	Approximated share of "new" employment	Approximated share of total employment Dec. 1987	Regional share commercial service employment SIC 6,7,8*
Metropolitan	60.1	55.9	51.9	42.2 (48.6)	51.1	44.8
"Spillover"	16.2	17.5	19.0	30.6 (20.1)	20.0	22.3
Brabant	5.1	7.2	9.3	12.6 (14.5)	9.9	12.2
Southern Peripheral	4.4	4.9	4.9	5.6 (6.4)	5.0	7.7
Northern Peripheral	14.2	14.5	14.5	8.8 (10.0)	14.0	13.0

Source: * Statistics of Employed Persons 1986. Approximations do not sum 100 because of rounding and establishments with "unknown" location

Table 10: Engineering in metropolitan areas

	Approximated share of employment in pre–1976 establishments	idem with respect to pre–1981 establishments	idem with respect to pre–1985 establishments	Approximated share of "new" employment	Regional share commercial service employment SIC 6,7,8*
Arnhem/Nijmegen	9.6	7.9	6.6	3.1 (3.6)	4.1
Utrecht	9.0	8.8	9.7	9.1 (10.5)	7.9
Amsterdam	14.3	11.8	10.0	8.6 (9.9)	14.0
The Hague	14.9	14.0	12.5	11.7 (13.5)	6.8
Rijnmond	12.3	13.4	13.0	9.7 (11.2)	12.1

Source: * Statistics of Employed Persons 1986. Approximations do not sum 100 because of rounding and establishments with "unknown" location

As the COROP region Delft attracted a few large engineering establishments in recent years, the "spillover" region scores relatively favourable on indicator 4. Consequently, in order to exclude this potential "bias", this indicator has also been re-estimated by excluding this region. These latter estimations (which imply an "over-estimation" of the performance of the other regions and an "under-estimation" of the "spillover" region) have been presented in parentheses in column 4.

Table 11: Space-time trajectory of "general information"

	Approximated share of employment in pre–1976 establishm.	idem with respect to pre–1981 establishm.	idem with respect to pre–1985 establishm.	Approximated share of "new" employment	Approximated share of total employment Dec. 1987	Regional share commercial service employment SIC 6,7,8*
Metropolitan	64.8	60.1	56.5	48.7	54.6	44.8
"Spillover"	16.8	18.0	19.6	23.4	20.6	22.3
Brabant	10.6	12.3	13.0	14.4	13.3	12.2
Southern Peripheral	2.5	3.1	3.5	3.1	3.5	7.7
Northern Peripheral	5.1	6.5	7.1	9.1	7.7	13.0

Source: * Statistics of Employed Persons 1986. Approximations do not sum 100 because of rounding and establishments with "unknown" location

Table 12: "General information" in metropolitan areas

	Approximated share of employment in pre–1976 establishments	idem with respect to pre–1981 establishments	idem with respect to pre–1985 establishments	Approximated share of "new" employment	Regional share commercial service employment SIC 6,7,8*
Arnhem/Nijmegen	1.1	2.2	2.5	5.0	4.1
Utrecht	7.9	8.8	9.4	12.0	7.9
Amsterdam	38.1	32.1	28.3	13.7	14.0
The Hague	5.2	6.2	6.2	8.5	6.8
Rijnmond	12.5	10.8	10.1	9.5	12.1

Source: *Statistics of Employed Persons 1986

References

Abernathy, W.J. and J.M. Utterback (1978)
Patterns of Industrial Innovation. Technology Review 80.

Andersson, A and B. Johansson (1984)
Knowledge Intensity and Product Cycles in Metropolitan Regions. Contributions to the Metropolitan Study 8, IIASA, Laxenburg.

Aydalot, P. (1984)
Reversals of Spatial Trends in French Industry since 1974. In: J.G. Lambooy (ed.) New Spatial Dynamics and Economic Crisis, Finnpublishers. Tampere, pp.41-61.

Batten, D. and B. Johansson (1987)
The Dynamics of Metropolitan Change. Geographical Analysis, 19, 3: 189-199.

Bearse, P.J. (1978)
On the Intra-Regional Diffusion of Business Service Activity. Regional Studies 12:563-578.

Berry, B.J.L. (1972)
Hierarchical Diffusion: The Basis of Developmental Filtering and Spread in a System of Growth Centers. In: N.M. Hansen (ed.): Growth Centers in Regional Economic Development. The Free Press, New York:108-138.

Bluestone, B. and B. Harrison (1982)
Deindustrialization of America. Basis Books, New York.

Brown, L.A. (1981)
Innovation Diffusion. A new Perspective. Methuen, London.

Buunk,J.B. and P.C.M. Elderman (1982)
De dynamiek van het bedrijvenbestand in Rijnmond.
Economisch Statistische Berichten: 1286-1290.

Centraal Bureau voor de Statistiek. Statistiek Werkzame Personen.
1973-1985. Voorburg.

Centraal Bureau voor de Statistiek, National Accounts. Several volumes, Voorburg.

Centrale Data Bank Woerden (1987)
Adressencatalogus 1987/88, Woerden.

Daniels, P.W. (1985)
Service Industries. Methuen, London.

Daniels, P.W. (1987)
Internationalization of Producer Services and Metropolitan Development.
Paper presented at the conference on "The Future of the Metropolitan Economy",
The Hague, 1987.

Davelaar, E.J. and P. Nijkamp (1987a)
Spatial Dispersion of Technological Innovation.
A Case Study for The Netherlands by means of Partial Least Squares.
Research Memorandum 1987-62, Department of Economics, Free University,
Amsterdam.

Davelaar, E.J. and P. Nijkamp (1987b)
Operational Models on Industrial Innovation and Spatial Development:
A Case Study For The Netherlands. Research Memorandum 1987-58.
Department of Economics, Free University, Amsterdam.

Davelaar E.J. and P. Nijkamp (1987c)
Innovative Behavior of Industrial Firms: Results from a Dutch Empirical Study.
Proceedings of the RICE conference Karlstad.

Davelaar E.J. and P. Nijkamp (1989)
Structural Transformation of Cities: The Case of The Netherlands.
In: H.J. Ewers and P. Nijkamp (eds.): Urban Innovation, forthcoming.

Duijn, J.J. van (1981)
Fluctuations in Innovations over Time. Futures 13: (4); 264-275.

Ewers, H.J. and P. Nijkamp (1990)
Sustainability of Urban Systems. Gower, Aldershot.

Freeman, C. (1987)
Technical Innovation, Diffusion, and Long Cycles of Economic Development. In: T.Vasko (ed.): The Long-Wave Debate. Springer, Berlin pp. 295- 309.

Freeman, C., J. Clark and L. Soete (1982)
Unemployment and Technical Innovation. A Study of Long Waves and Economic Development. Frances Pinter, London.

Granstrand, O. (1986)
The Modelling of Buyer/Seller-Diffusion Processes. A Novel Approach to Modelling Diffusion and Simple Evolution of Market Structure. Paper prepared for the conference on "Innovation Diffusion", Venice.

Haan, M.A. de and P. Tordoir (1986)
Het belang van diensten voor de Nederlandse economie. Gangbare en nieuwe inzichten. PSC-TNO, Delft.

Hirschman, A.O. (1958)
The Strategy of Economic Development. Yale University Press, New Haven.

Howells, J.R.L. (1983)
The Location of Research and Development: Some Observations and Evidence from Britain. Regional Studies 18: 13-29.

Holland, S. (1976)
Capital versus the Regions. MacMillan, London.

Huppes, T. (1987)
Over netwerken en organisaties, Economisch Statistische Berichten: 932-940.

Keeble, D., P.L. Owens and C. Thompson (1983)
The Urban-Rural Manufacturing Shift in the European Community. Urban Studies 20: 405-418.

Keeble, D. and E. Wever (1986)
New Firms and Regional Development in Europe. Croom Helm, London.

Kleinknecht, A. (1987)
Innovation Patterns in Crisis and Prosperity. MacMillan, London.

Koerhuis, H. and W. Cnossen (1982)
De software-en computerservice bedrijven. GIRUG, Groningen.

Kuznets, S. (1930)
Secular Movements in Production and Prices. Houghton Mifflin, Boston.

Lambooy, J.G. and N. Tates (1983)
Zakelijke diensten, een complementaire sector. Economisch Statistische Berichten: 676-680.

Lambooy, J.G. and P. Tordoir (1985)
Professional Services and Regional Development. Fast-paper no.74, European Commission, Brussel.

Leone, R.A. and R. Struyck (1976)
The Incubator Hypothesis: Evidence from five SMSAs. Urban Studies 13: 325-331.

Malecki, E.J. and P. Nijkamp (1988)
Technology and Regional Development: Some Thoughts on Policy. Environment and Planning C 6: 383-399.

Mensch, G. (1981)
Stalemate in Technology: Innovations Overcome the Depression. Ballinger, New York.

Metcalfe, J. (1981)
Impulse and Diffusion in the Study of Technical Change. Futures 13 (5): 347-359.

Molle, W.T.M. and J.G. Vianen (1982)
Het vertrek van bedrijven uit Amsterdam. Economisch Statistische Berichten: 608-613.

Moss, M.L. (1985)
Telecommunications and the Future of Cities. Paper presented at the Conference on Landtronics, London.

Naisbitt, J.(1984)
Megatrends. Warner Books, New York.

Nelson, R.R. and S.G. Winter (1982)
An Evolutionary Theory of Economic Changes. Harvard University Press, Cambridge, Mass.

Nijkamp, P. (ed) (1986)
Technological, Employment and Spatial Dynamics. Springer, Berlin.

Noyelle, T.J. and T.M. Stanback (1984)
The Economic Transformation of American Cities. Rowman and Allanheld, Totowa, New Jersey.

Oakey, R.P., A.T. Thwaites and P.A. Nash (1980)
The Regional Distribution of Innovative Manufacturing Establishments in Britain. Regional Studies 14: 235-253.

Potters, A.L.M. (1983)
Dienstverlening in beweging. Stedebouw en Volkshuisvesting: 169-176.

Pred, A. (1977)
City System in Advanced Economies. Hutchinson, London.

Rosenberg, N. (1976)
Perspectives on Technology. Cambridge University Press, Cambridge, Mass.

Rothwell, R. and W. Zegveld (1985)
Reindustrialization and Technology. Longman, Essex.

Sahal, D. (1980)
The Network and Significance of Technological Cycles.
International Journal of Systems Science 11 (8): 985-1000.

Schmookler, J. (1966)
Invention and Economic Growth. Harvard University Press, Cambridge, Mass.

SKIM (1981)
Microscope for Economies, Rotterdam.

Thompson, W.R. (1970)
Internal and External Factors in the Development of Urban Economies.
In: H.S. Perloff and L. Wingo (eds.): Issues in Urban Economics.
The John Hopkins Press, Baltimore, pp. 43-62.

Thwaites, A.T. (1978)
Technological Change, Mobile Plants and Regional Development.
Regional Studies 12: 445-461.

Vasko, T. (1987)
The Long-Wave Debate. Springer, Berlin.

Vernon, R. (1966)
International Investment and Institutional Trade in the Product Cycle.
Quarterly Journal of Economics 80: 190-207.

Wever, E. (1984)
Nieuwe ondernemingen in Nederland. Van Gorcum, Assen.

Zwan, A. van der (1979)
On the Assessment of the Kondratiev Cycle and related Issues.
Centre for Research in Business Economics, Erasmus University Rotterdam (mimeo).

3.3 Promoting and Inhibiting Factors in the Regional Environmental System

Edward J. Malecki

The regional environment for innovation must be understood in the broader context within which innovation itself is presently seen (Dosi 1988; Kline and Rosenberg 1986). The traditional linear model of technological change begins with formal R&D, and basic research in particular, and expects that innovations and higher productivity will result, but nationally and, by extension, locally. Despite considerable evidence that such a linear process rarely occurs, and certainly not without complex interactions among activities (Buswell 1983; Kline and Rosenberg 1986), R&D remains an important means of searching for new technical knowledge and is a measure of the input to technological change at the national level, although perhaps not at the level of the firm or the region.

However, R&D is costly and poses barriers to entry for new and small firms, and this has regional dimensions. Important informal processes, not captured by R&D statistics, take place through the monitoring of information and technological capabilities, through publications, technical associations, watch-and-learn processes, and personnel mobility. Learning by doing and learning by using provide major opportunities for technological improvements and represent 'externalities, internalized within each firm' (Dosi 1988: 1125; Dutton and Thomas 1985). A fourth process is the adoption of innovations originating in other industries and embodied in capital equipment and intermediate inputs (Dosi 1988; Pavitt 1984). Taken together, these mechanisms broaden considerably the scope of information-gathering activities required of firms. As informal activities, they are difficult to measure, if not to identify, and they are to a large extent embodied in people and in organizations.

The extent to which innovation depends on people, their accumulated knowledge and capabilities gained through experience, and the information and contact networks on which they draw, is a striking characteristic of recent research (Dosi 1988; Malecki 1989).

Much of this knowledge is tacit rather than public and frequently consists of the know-how needed to link distinct bodies of technology. This type of synthesis of technologies is perhaps more important than previously as microelectronics and computer software have become integral to the processes of many other industries. Taken together, these tendencies and capabilities regarding technical knowledge define a technological trajectory, which is fairly stable, although evolutionary, over time. It takes into account economic as well as technological constraints, but serves as a path outside of which innovation is rather less likely at any given time (Dosi 1988).

1 The Regional Environment for Innovation and Entrepreneurship

A regional setting or environment likewise has its fixed skills and capabilities, although the trajectories themselves are defined by the prevailing technological paradigm, as embodied in its population of people and of firms. Even in the absence of interfirm linkages, there is a "collective asset" of groups of firms and industries within countries or regions that represent technological externalities (Dosi 1988:1146). To a large degree, these are the synergies referred to less explicitly by Anderson (1985) and Stöhr (1986) in the context of creative regions and territorial innovation complexes. (Dosi's [1988:1147] recognition of spatial and regional dimensions of technological capabilities is also quite significant within mainstream economics.) Obviously, the synergies as well as the direct linkages among firms will vary according to economic sector and the ways in which innovative activities proceed in various sectors. For example, Pavitt (1984) categorizes industries into four types, according to their dominant source of technological innovation: supplier-dominated, specialized suppliers, scale-intensive sectors, and science-based sectors. These correspond to the ways in which, and where, knowledge is accumulated and utilized in innovation.

To a large degree, the environments where innovation flourishes and where new firm formation takes place at a high rate are dependent on technological capability even outside "'high tech" sectors. It is clear that the location factors which predominate in "high tech" industries and in the majority of innovation-oriented regional policies are broader than those common in previous generations of regional policies. The importance of information and instability in the process of technological change itself brings this about (Molle 1983). However, the recent attempts to bring about more widespread growth are perhaps not very different from earlier growth pole and growth center policies (Oakey 1984). The most ambitious example of deliberate high technology development is the Technopolis Concept in Japan, a plan to build a network of 19 regional high tech cities linked to Tokyo by bullet trains (Smilor et al. 1988; Tatsuno 1986). The criteria used by MITI (the Ministry of International Trade and Industry) to select the locations include: proximity to a "mother city" of at least 150,000 to provide urban services, proximity to an airport or bullet train station, an integrated complex of industrial, academic, and residential areas, and a pleasant living environment. These are thought to be positive elements that will ensure the success of these Japanese "silicon valleys." Much emphasis is also placed on high levels of local amenities for the attraction and retention of mobile technical workers (Keeble and Kelly 1986). Labor was traditionally undifferentiated in regional economic research except perhaps by cost, whereas now it is recognized that quite a distinct labor market operates for professional and technical workers from that for production employees (Hall and Markusen 1985). The availability of technical and professional workers is related to technological agglomeration in a complex way (Malecki 1987). The mobility of these workers and their access to information about employment in a large number of possible locations make it particularly difficult to foresee any reduction in their tendency to agglomerate in space. For the workers themselves, large urban regions maximize their alternative employment opportunities within the range of daily commuting distance. In order to maximize the likelihood of obtaining a sufficient number of professional workers, firms in turn locate in large cities. The factors which influence the location of R&D facilities, for example, are associated much more with large urban areas than are the factors important to manufacturing (Table 1). Professional and technical labor, air transportation, and quality of life are variables that have reinforced the importance of urban regions as innovative environments.

Not all urban regions are the same, of course, and among the most critical variations among places are information, contacts, and innovativeness. Information and contact networks encompass both publicly available (albeit perhaps costly) information and scientific knowledge and technology as a set of abilities and skills embodied in people. Geographical variations are inherent in those human elements. "Innovative activities present - to different degrees - firm-specific, local and cumulative features" (Dosi 1988: 1131). Although people and their knowledge are mobile, and become key elements in technology transfer and entrepreneurship, there is a great deal of stability over time in the firms and locales where innovative ability has accumulated.

2 Regional Factors which Promote Innovation and Entrepreneurship

Many lists are available of factors that seem to have positive effects on innovation and new firm formation at the regional level (Thomas 1988; Shapero 1984; Jowitt 1988;

Table 1: Location factors influencing manufacturing plants, R&D facilities, and new start-up firms

Manufacturing Plants
- Labor availability and costs
- Highway and rail transportation
- Accessibility of raw materials
- Business climate

R&D facilities
- Availability of professional talent
- Air transportation
- Quality of life
- Highway transportation

New start-up firms
- Seed capital
- Entrepreneurial climate
- Venture capital
- Incubator space
- Bank financing
- Space for expansion

Sources: Browning (1980); Gruenstein (1949)

Miller and Coté 1987; Senker 1985). Some of these are by now well known, such as Bruno and Tybejee's (1982) twelve factors "essential" for "the environment for entrepreneurship" in previous research. These are, in apparently ranked order:

1. Venture capital availability
2. Presence of experienced entrepreneurs
3. Technically skilled labor force
4. Accessibility of suppliers
5. Accessibility of customers or new markets
6. Favorable governmental policies
7. Proximity of universities
8. Availability of land or facilities
9. Accessibility to transportation
10. Receptive population
11. Availability of supporting services
12. Attractive living conditions

Shapero (1984) distinguished between personal and environmental factors associated with new firm formation:

1. Personal characteristics
   - Education
   - Independence
   - Willingness to accept risk
   - Prior experience in small firms
2. Environmental factors
   - Role models
   - Successful experiences of others
   - Population of small businesses
   - Support of local financial community
   - Entrepreneurial climate
   - Innovative technical information
   - Business information

Our knowledge of the nature of this environment is still sketchy, but it is apparent that some regional environments are more stimulating to innovation and entrepreneurship than others. For example, the mix of industries and of sizes of firms in a region are factors long recognized as influencing the rate of firm formation (Beesley and Hamilton 1986; CURDS 1979; Sweeney 1987). In addition, proximity to urban areas and their bundle of agglomeration advantages provides benefits not available or less abundant in remote milieux. Innovation and entrepreneurship, as nonroutine economic activities, display a strong tendency toward agglomeration (Malecki 1987; Oakey 1985). The information-intensive nature of technological activities and the resultant need for face-to-face communication favor those places that offer:

1. High levels of competence
2. Many fields of academic and cultural activity
3. Excellent possibilities for internal and external communications
4. Widely shared perceptions of unsatisfied needs
5. A general situation of structural instability facilitating a synergistic development (Andersson 1985).

These conditions for "regional creativity" can be translated into more conventional policy variables, but in general, they focus on three main elements: (1) the presence of professional and technical labor (competence); (2) urban agglomeration, or a threshold size of place, where cultural activity and communication will be heightened; and (3) conditions that promote synergy or instability (Stöhr 1986; see also CURDS 1979: 123).

Venture capital is one of the most difficult regional factors to deal with. It appears in virtually every inventory of "necessary" conditions for innovative entrepreneurship, and is a high priority for policy initiatives in Western countries (Houttuin 1985; OECD 1985). It must be stressed that venture capital, particularly in the American context, is not the same as loan funds, and there is little such funding from public-sector capital programs, which usually lend money and expect a full and regular payback. Shapero (1971) emphasized the significance of the local financial community on the potential for local entrepreneurship. The appearance of the first, "almost random" company formations in an area are the most difficult to account for, and may not always precipitate any follow-up of further entrepreneurship. He found substantial differences among localities in the degree to which local banks were willing to lend for new, untried ventures. If an area attained a level of sustained entrepreneurship, it was typically associated with the growth of a financial, legal and service community to support it.

The supply of capital for firms attempting to start in business varies greatly from region to region (Florida and Kenney 1988). However, it is not simply the "availability" of capital in a formal sense that significantly promotes firm formation. In the USA at least, there are networks in some places of individuals, many of whom are former entrepreneurs, who are willing to provide start-up capital, often informally, to new businesses (Thomas 1988; Gruenstein 1984; Miller and Coté 1987; Wetzel 1986). This "entrepreneurial climate" of a place is probably the single most important variable influencing entrepreneurship as an ongoing part of urban life. This climate relies almost entirely on a well-connected network of investors, especially informal investors, previous entrepreneurs, and an aura of nonroutine, innovative activity (Gruenstein 1984; Miller and Coté 1985; Senker 1985; Shapero 1984; Sweeney 1985). The entrepreneurial climate of a region also is affected by the power elite in a community (Schell 1983; 1984). The interaction between public and private sector in a locale, indeed, appears to be of considerable importance in creating and maintaining contacts, and in creating a supportive milieu for new firms (Dyson 1988; Fosler 1988; Miller and Coté 1987). Favorable government policy, a receptive population, and attractive living conditions have a major influence on entrepreneurial activity in an area. Most important to the "entrepreneurial climate" of a region, however, are entrepreneurial "role models." These can promote and encourage local entrepreneurship and make seed and early financing available to new entrepreneurs (Schell 1983 1984; Gruenstein 1984). Financing for start-up firms can come from many sources, including personal funds, family and friends, local bankers, and outside lenders, but it tends to come from informal (noninstitutional) sources that operate almost entirely via a network of personal and local contacts (Wetzel 1986). In addition to their role in financing new firms, previous entrepreneurs share a variety of routine business knowledge that is necessary to operate a business, such as regulations, taxes, accounting, suppliers, customers, and marketing and distribution.

Others suggest that a favorable environment also involves a cultural milieu that accepts change, acknowledges entrepreneurship, has a relatively uninhibited flow of ideas and information from outside the region, and an awareness of "best practice" in the industry. An outward-looking attitude, a constellation of suppliers and infrastructures particular to the production requirements of a group of firms also supports the needs of new firms. In addition, a high level of technical skill in the local industry or industries is essential. A threshold size of region appears to be essential, but rural entrepreneurship in Denmark and other countries shows that this is not necessarily the case (Illeris 1986).

Communication and contact networks of key individuals are key to understanding the fundamental role of metropolitan regions in the creative process. The contact networks of technical workers are a principal source of technical knowledge. Firms rely on the heterogeneous contacts and networks of individuals in order to acquire information from other organizations. The information "gatekeeper" role is a well-known one; only recently has the diversity of contacts of technical workers been studied (Hamfelt and Lindberg 1987). Increasingly, it is seen to be vital for firms to recruit personnel with certain contacts and personal networks in place, to encourage contacts to be formed and maintained, and to circulate people in order to more fully develop their networks of contacts to the advantage of the firm (Hükanson 1987; Shapero 1985). The common contact networks of rival firms even results in informal trading of know-how between competitors (von Hippel 1988).

Andersson's conditions for regional creativity can be translated into more conventional policy variables, but in general, they focus on three main elements: (1) the presence of professional and technical labor (competence); (2) urban agglomeration, or a threshold size of place, where cultural activity and communication will be heightened; and (3) conditions that promote synergy or instability. Entrepreneurship follows from the technological capability that develops in such settings, and is enhanced by communication of new ideas and knowledge among technical people. Thus the local labor market and its relationship to larger (national and international) markets is important. The selectivity of professional and technical workers in where they live and work is an important constraint on which places can actually retain the specific labor force necessary for a vibrant economy (Malecki 1987).

2.1 Best-Practice and Regional Competitiveness

Technological capability in firms or in regions is not fixed or permanent, since technology and the abilities of competitors are constantly changing (Dosi 1984; Nelson and Winter 1982). It relies on the activities of firms, and how close those activities are to the state-of-the-art (best-practice) and international standards of product design and quality at any point in time. "Best practice" technology typically refers to process technology (machinery, equipment, management practices) which ranges from that of the most efficient (best-practice) producers to the most inefficient (worst-practice) (LeHeron 1973; Malecki 1983: 93-94). The spectrum from best-practice to worst-practice technology is largely a function of the age or vintage of capital equipment employed. Newer vintages will incorporate or embody newer concepts, techniques, and knowledge which tend to give an advantage to firms - and regions - where newer technology is employed (Varaiya and Wiseman 1981). Firms and countries which are not near the current "technology frontier" both in science and in production find it increasingly difficult to keep up with changes in other places (Cohen and Zysman 1987; Spence and Hazard 1988).

A concern for regional innovative capability became a focal point in regional research a few years ago, but it was concerned almost equally with large and small firms: the overall technological level of the region was the focus (CURDS 1979; Ewers and Wettmann 1980; Thwaites 1982). However, the predominant empirical finding about small firms in peripheral areas is their relatively low level of R&D and innovativeness, compared to those located on the fringe of major urban regions, both in the Netherlands (Pellenbarg and Kok 1985), and in more densely populated regions containing urban agglomerations in the West Germany (Meyer-Krahmer 1985). The concentration of R&D in the UK is even more biased against peripheral regions (Goddard and Thwaites 1986).

The level of technology in a region, however, goes beyond considerations of machinery and plant design. It encompasses the stock of knowledge within a firm or region concerning both what is being made and how things are made. Technically progressive firms obtain knowledge from customers and suppliers as well as generate it internally. For a region, links to outside information networks and a high level of technical skill define a level of best-practice that in many respects is cultural and which goes well beyond what is encompassed by R&D (Brugger and Stuckey 1987; Sweeney 1987). Most of all, then, technological capability relies on people in technological roles within organizations. Not all people will be able to serve as "gatekeepers" but their role is pivotal. For regions, the gatekeeper function also is critical, if informal. Sweeney's (1987) suggestion that "extension agents," much like those who have helped to diffuse agricultural technology around the world, can serve this purpose. Britton (1988) suggests a similar "active" liaison, which would connect firms with the appropriate consultants (located in the region or otherwise) who are familiar with best-practice products and processes.

Sweeney (1987) believes that technology transfer depends critically on the level of information that is "in the air" in a locale (Andersson 1985; Rogers 1982). This quantity of information hinges on the firm size mix in the region: small firms much more willingly share information with each other, largely because this open flow has mutual benefits (Russo 1985; Brusco 1986). Large firms are more likely to have to devote much of their information-oriented time to intrafirm communications, such as between any facility and the firm's headquarters or between facilities. This is particularly true of formal R&D, for which the imperatives of interaction with both headquarters and with production plants and sales offices are large.

Openness of a region to new ideas may by measurable by the diversity of imports (Johansson and Westin 1987). Regions where imports are diverse and state-of-the-art are most likely to be innovative themselves.

The critical element in the process is the "innovation response to import flows." If the import node and its surrounding region of smaller nodes have a sufficiently rich production structure embedded in an active R&D environment, with access to a creative and knowledge-oriented labor force, then this together with the import of new products constitutes a prerequisite for successful innovations. The latter may range from new products in the form of imitations and modifications of imported products adapted to local demand to the creation of products which are complementary to developing import flows. When this type of response is successful, it brings about the birth of local product cycles (Johansson and Westin 1987: 15).

Some of the environment of a region also depends on the legal and institutional structure within which firms must operate. Teece (1986) provides a valuable typology of the "regimes of appropriability," those environmental factors, other than firm and market structure, that govern an innovator's ability to capture the profits generated by an innovation. The most important dimensions of such a regime are the nature of the technology and the strength of legal mechanisms of protection. Giaoutzi (1985) points out how these mechanisms work to the disadvantage of less-developed countries in particular. Patents, copyrights, and trade secrets (such as chemical formulas) are legal means of keeping a technology out of competitors' hands, but these do not guarantee that imitation cannot take place.

Most common is the evolutionary transformation of products and processes over time. In the early, "preparadigmatic" stage of a technology's development, an innovation can become worthless quickly as new designs leapfrogs it. As standardization of design takes place, an industry paradigm emerges, and the firm which gains most will tend to be the one which has a full array of complementary assets in order to exploit not only the innovative design but also to manufacture it competitively and bring it successfully to market. These assets include competitive manufacturing, distribution, service, a successful trade name, and complementary technologies (Teece 1986: 289). Traditional neoclassical trade theory refers to regimes of "tight" appropriability and zero transactions cost, where it does not matter whether an innovating firm has an in-house manufacturing capability, domestic or foreign. It can simply engage in arms-length contracting (such as licensing and co-production) for the sale of the output of the activity (R&D) in which it has a comparative advantage. In a regime of "weak" appropriability, manufacturing may be necessary if an innovator is to appropriate the rents from an innovation. This is especially true where the requisite manufacturing assets are specialized to the innovation. If an innovator's manufacturing costs are higher than those of imitators, for example, the bulk of the profits will go to the imitators.

2.2 Entrepreneurship and Technological Capability

Entrepreneurship is a form of technology transfer, where knowledge is taken from one firm to a newly founded one by an individual. Coffey and Polése (1985) consider entrepreneurship to be at the core of the process of local economic development, especially in light of experience with earlier regional policies whose principal consequence may have been "external control" (Watts 1981). Information channels are central to the transfer of knowledge from person to person, from firm to firm, and from region to region.

Allen et al. (1983) found that information among firms in Ireland, Spain, and Mexico flowed mainly through informal channels - that is, through interpersonal contacts. The channels are more formal among large firms and organizations. As Sweeney (1985; 1987) suggests, foreign subsidiaries obtained the greatest proportion of their technology from their parent firms. In fact, such plants had several channels of technology blocked to them, such as government-sponsored research, trade fairs, and industry associations, which are quite important to domestic firms. Firms in the same industry are also much more likely to be a source of technological knowledge to domestic firms. Allen et al. conclude that domestic firms in many ways have easier access to foreign technology than do the subsidiaries of multinational firms.

Orientation toward national and international markets is what distinguishes among types of small firms. Brusco (1986) suggests three types: the traditional artisan common to many rural areas; dependent subcontractors who produce parts and components to be included in products of larger firms which sell nationally and internationally; and small firms in industrial districts. The technological sophistication of the latter group of firms must be quite great, often for a large number of clients, which indicates that their level of quality and technical standards are high. In industrial districts, an additional type of firm is found: producers of machinery necessary for the production of the district's specialized commodity. In industrial districts, "the local communities have accumulated managerial, technical and commercial competence and capacity... This knowledge is not characteristic of one particular group but...is spread to all social strata" (Brusco 1986: 198). Both Russo's (1985) case study of the ceramic tile industry and Sabel et al.'s (1987) example of German textile-machinery firms illustrate that the interaction between machinery firms and their users generates a continual flow of information concerning improvements and modifications. In other words, information based on "learning by using" flows back and forth between the two industries (Kline and Rosenberg 1986; Rosenberg 1982). Indeed, Lorenzoni and Ornati (1988) suggest that firms in such "constellations" are more willing to seek information from outside sources, such as consultants, universities, and other firms. The "environmental texture" of such districts-suppliers and infrastructures particular to the production requirements of a group of firms-supports the needs of new firms.

New firm formation, which also depends on flows of knowledge, varies geographically in part because of differences in innovativeness and in whether and how innovativeness is promoted. Birch (1987: 140-165) lists five factors that enhance entrepreneurship, that are high quality in some places and not in others:

1. Educational resources, particularly higher education
2. Quality of labor
3. Quality of government
4. Telecommunications
5. Quality of life

Birch's designation of educational resources is actually much more specific than this; he insists that research-based universities, at the leading edge of change, are the necessary ingredient. Labor quality refers to skilled and adaptable workers, for whom training and adjustments are less costly, regardless of the wage level. Education, at the primary and secondary levels (through age 18), is often a key dimension of this adaptable workforce. Quality of government refers to the balance between public services and their cost-"tax-efficient" in Birch's terms.

"Telecommunications are to the high-innovation economy what railroads, rivers, canals, and highways were and are to the industrial world." Innovation-based firms generally do not have much in the way of physical objects to move around, and their raw materials are few, compact, and easily transported. Instead, most of what they move about is information - the kind contained on credit card slips, airline reservations, bank transactions, insurance claims, securities sales, personnel records, commercial credit reports, and mail orders. "Locales that offer state-of-the-art communications facilities have a great advantage in this kind of economy" (Birch 1987: 146-147; Cohen and Zysman 1987).

Birch recognizes that quality of life is strongly associated with urban size. "Whatever it is, the quality of life has to do with "interesting pastimes" and an enjoyable atmosphere after work. Whatever form it takes, "nice, exciting" places have decided advantages when it comes to appealing to executives of innovative firms for these are the areas where entrepreneurs and employees will want to settle and raise their families" (Birch 1987: 147-148).

The various conditions discussed thus far can also be classified into regional infrastructure, spatial concentration, and quality of life (Thomas 1988). In Thomas's account, infrastructure includes capital, the economic-industrial base of technologically adept firms whose inputs and skills are transferable to new firms, a quality labor supply, technology networks, transportation and communication systems, and a socio-political structure of supporting institutions. Spatial concentration refers to the availability of input-output linkages and the synergy present in agglomerations. Quality of life is a less precise element that appears to operate through labor market choices of professional, especially R&D, workers (Keeble and Kelly 1986; Malecki 1987). These comprise the elusive "regional factor" which has proved difficult to define cxactly (Thomas 1987). However, local skills and entrepreneurship, unlike scientific knowledge technology, are more local phenomena, not readily transferable from place to place, and they depend on local economic structure as well as on the attitudes and culture of the region (Brugger and Stuckey 1987; Sweeney 1987).

3 Regional Factors Inhibiting Innovation and Entrepreneurship

A list of the regional characteristics that constrain or inhibit innovation and new firm formation can be constructed merely by listing the absence or relative shortage of the ingredients that encourage entrepreneurship and innovation. Often, discussions of the topic end with that observation. Inhibiting factors include an unfavorable industrial mix, domination by branch plants of large firms, restrictions on imitation by other firms, a concentration of jobs lacking in skills, low levels of contact with outside networks, and, generally, a lack of urban agglomeration advantages, such as suppliers, markets, information, infrastructure and capital. These shortcomings are greatest in rural regions, where urban characteristics are normally lacking by definition, and in peripheral regions, located outside a nation's technological heartland. The constraints on innovation and entrepreneurship are great but not equal in all such places (Malecki 1988a; Reid 1988). The emergence of new firms depends most "on community and business leaders who are informed about current practices and are willing to help newcomers in the commercialization and improvement of crafts, skills, and industries in the local economy. This support will not emerge in all communities" (Malecki 1988a: 23).

Other constraints relate to the socio-economic mix common to rural areas, including personal characteristics noted by Shapero (1984). The most notable problem in many regions, urban as well as rural, is low levels of education. This is associated, in turn, with low levels of skills and business knowledge and a reduced range of contacts and information sources (Miller et al. 1986). "Good" entrepreneurs are educated, experienced in a variety of places and business settings, and have professional contact networks in place (Keeble and Wever 1986). Other cultural influences on entrepreneurship include the entrepreneurial culture of a locality - the examples of others, an already strong small firm sector, the tradition of being self-employed. If lacking, as is the case in many rural or peripheral areas, innovative growth and entrepreneurship are unlikely (Illeris 1986; Sweeney 1985: 91). Some rural areas are burdened by other cultural/social factors, such as no established entrepreneurial class, a social structure dominated by church or a single industry (such as mining), and high rates of out-migration. However, many of these shortcomings can be overcome if education levels among the population are high; education is strongly related to a greater range of information sources and a wider set of customers and markets - in short, to an awareness of industry best-practice (Ewers 1986; Miller et al. 1986).

In the context of most small, peripheral regions, however, innovative development may be "too good to be true" (Martin 1986: 17; Brugger 1986: 42; Friedmann 1986). The human capital of such regions is too limited and too mobile, and many of the talented people will simply leave. Historically, indeed, out-migration selects the most productive subset of the population of rural regions. Clearly, the inhibiting factors discussed thus far are relevant in the context of less developed countries as well, and there is much that can be learned from the experience of those "even more peripheral" regions (Dahlmann and Westphal 1982; Fransman and King 1984; Giaoutzi 1985; Malecki 1988b; Segal 1987).

Unless a local economy meets some fairly large threshold size, its base of potential entrepreneurs - and the likelihood that they as a group will be able to come up with successive rounds of innovations as the product cycle progresses - will be inadequate to compete with other regions (Martin 1986). The issue of threshold size is certainly central in most accounts of supportive regional environments (Andersson 1985; Malecki 1987; Oakey 1985), but it can be overemphasized. An array of other influences may be more important than urban size alone.

Among the factors inhibiting the indigenous potential of peripheral regions, most attention has focused on the dependence of such regions on branch plants of large, multinational corporations. The evidence is clear that such plants tend to have lower levels of highly-trained and skilled personnel, certainly lower levels of R&D, and little attention to nonroutine activities or new products (CURDS 1979; Goddard and Thwaites 1986; Malecki 1981; Miller and Coté 1987; Sweeney 1987; Watts 1981). Regions dominated by large firms and by branch production plants are unlikely to have the level of information, of R&D, and knowledge of the state-of-the-art needed to spawn new firms. This is summed up well by Sweeney (1985: 97):

> [A]reas dominated by large firms tend to have low entrepreneurial vitality. Such firms have internalised their information resources and networks. ... Areas dominated by branch plants of large firms also have low entrepreneurial vitality, their networks are with their distant parent company. Sub-contracting, an important source of start-up for new companies, is minimised by the large integrated firm and even more by branch plants. An important means of developing local networks for technology and information transfer is thus lost.

Localities which are dominated by large firms are not centers of entrepreneurial vitality, partly because there is a lack of openness in the flow of information from the large to the small. Large bureaucracies prefer to talk to large bureaucracies (Sweeney 1985: 92-94).

Coffey and Polése (1985) propose four stages of local development to counteract large-firm dominance in peripheral regions:

1. The emergence of local entrepreneurship
2. The growth and expansion of local enterprises (to replace jobs lost by closed branch plants)
3. The maintenance of local enterprises under local control
4. The attainment of an autonomous local control structure and of a local business service sector

Whether or not such a process of local development can occur depends on local conditions. "Entrepreneurial vitality is very much a local phenomenon. Prosperity and economic growth of regions and localities are strongly associated with the strength and vitality of the small firm sector in the region or locality" (Sweeney 1985: 94; see also Beesley and Hamilton 1986). It must be stressed that new firms are rarely attracted from elsewhere; entrepreneurship either is a standard endeavor in a region or it is not. Cooper (1985) found that nearly all (84%) of firms are founded within commuting distance of where the founder previously worked.

Diversity within the small firm sector, not a dependence on a narrow range of technologies or markets, reinforces this vitality. "In other words, localities having high entrepreneurial vitality have characteristics similar to those of technically progressive firms, openness in giving as well as taking information, continuing effort in their own search for information and good internal communication flows" (Sweeney 1985: 95-96). Entrepreneurial vitality, once initiated, becomes self-reinforcing and sustaining. It also is a key element in successful exchange of research information from R&D to users (Segal 1987). In such a contact system, users also become important sources of innovation, both of product designs and of production processes (Von Hippel 1988). This "information rich environment" (Andersson 1985: Johansson 1987) and "technical effervescence" (Miller and Coté 1987) are the principal characteristics of "creative regions." Local information channels and knowledge are fundamental to entrepreneurs.

Large firms try to maximize information-intensive contacts by centralizing in large urban centers and around locales with innovative small firms (Pottier 1988). Around these cluster service activities for whom large firms are the dominant market. In addition to new firm formation, some regions have a greater proportion of successful, more rapidly growing small firms than other regions. Again, social and occupational influences appear to be most significant in reinforcing existing spatial contrasts. Mason (1985) found the bulk of the most technologically and financially successful firms in the UK in the prosperous South East. How do such regional variations come about? In a recent Dutch study, successful entrepreneurs (as opposed to those of failed or failing firms) were more prepared for starting a business: they had more starting capital, were more oriented to markets outside the local region, had a larger number of clients, and were somewhat more likely to have worked for a large firm previously (Wever 1986). Keeble and Kelly (1986) also found that virtually all of the founders of new computer electronics firms had a higher education, and 80% had a degree. Overall, there is a strong tendency for entrepreneurship to be strongest precisely in those regions which need it least, suggesting that to rely on new and small firms will not eliminate regional economic differentials.

It is clear from lists of regional environmental factors behind "creative regions" that several of the factors are simply conditions common to most, if not all, large urban regions (Andersson 1985; Malecki 1987; Malecki and Nijkamp, forthcoming). For example, availability of land and facilities, accessibility to transportation, and attractive living conditions tend to be attributes found in virtually any major metropolitan region, with the exception of some older industrial cities. Pottier (1985) believes that advanced technology is possible only in large urban regions where this agglomeration of attributes is concentrated. It is the other, human factors that appear to vary most among regions.

The "right" environment may exist in large urban regions (and nearly always the suburban or fringe areas of such regions), or in small towns in attractive settings, but it tends to be within reach of major airports and other essential urban amenities (Keeble and Wever 1986). A high rate of new firm formation reflects the information networks of professional and technical workers and their ability to identify sources of funding (especially venture capital) for the establishment of new firms. The "spin-off" process, frequently identified as a consequence of R&D in a regional context, is a subset of the general entrepreneurial process which takes place in facilitating (usually urban) environments. New firms in high technology sectors are even more dependent than firms in other sectors on the agglomeration of technical workers and on city size generally (Armington et al. 1984).

Conclusion

In conclusion, there may be relatively little that policies as typically viewed can do to alter the situation (Goddard and Thwaites 1986; Malecki and Nijkamp, forthcoming). Some possibility exists that a small region can become a technological complex, or that a declining city can revitalize itself around new technologies. More often, however, the choice must be among large, already rather prosperous urban regions (Pottier 1985). The promotion of entrepreneurship on a regional basis, as has been tried notably in the Federal Republic of Germany, is too recent to be considered a long-term success. At the least, it is clear that R&D alone is insufficient to spawn creativity and firm formation on a widespread basis (Cooke 1985; Malecki and Nijkamp, forthcoming).

Birch (1987) concludes that the environment for high-innovation start-ups cannot be created overnight - or even in a generation. Cultural centers and research universities can be created, but this takes time. “Much can be done, but it takes will - and a willingness to raise taxes to pay for the amenities” (Birch 1987: 163). The long-term nature of developing a supportive regional environment is less frequently acknowledged, but it is clearly central to the innovative and entrepreneurial process in creative regions. Note, however, that Birch says little about the human contacts and knowledge flows currently stressed by researchers with a regional orientation.

In particular, policies cannot create entrepreneurial role models, and venture capital is different from conventional subsidies. Most important is a long-term view where innovativeness is routine, education and information exchange are the local norm, and successive generations of innovators and entrepreneurs are nurtured by their predecessors. Public sector policies alone cannot create this environment, but attitudes and practices can be sparked by sincere and persistent efforts by both government and business leaders.

References

Allen, T.J., Hyman, D.B. and Pinckney, D.L. (1983)
Transferring technology to the small manufacturing firm:
a study of technology transfer in three countries. Research Policy 12: 199-211.

Andersson, A.E. (1985)
Creativity and regional development.
Papers of the Regional Science Association 56: 5-20.

Armington, C., C. Harris and M. Odle (1984)
Formation and growth in high-technology firms: a regional assessment.
In: Technology, Innovation, and Regional Economic Development pp. 108-144.
Washington, DC: US Government Printing Office.

Beesley, M.E. and Hamilton, R.T. (1986)
Births and deaths of manufacturing firms in the Scottish regions. Regional Studies 20: 281-288.

Birch, D.L. (1987)
Job creation in America. New York: The Free Press.

Britton, J.N.H. (1988)
Regional innovation policy. Paper presented at the North American Meetings of the Regional Science Association, Toronto, November 1988.

Browning, J. (1980)
How to select a business site. New York: McGraw Hill.

Brugger, E.A. (1986)
Endogenous development: a concept between utopia and reality. In: Self-reliant development in Europe, M. Bassand et al., eds., pp. 38-58. Aldershot: Gower.

Brugger, E.A. and Stuckey, B. (1987)
Regional economic structure and innovative behaviour in Switzerland. Regional Studies 21: 241-254.

Bruno, A.V. and T.T. Tybejee (1982)
The environment for entrepreneurship. In: Encyclopedia of entrepreneurship, K. Vesper, ed., pp. 288-307. Englewood Cliffs, NJ: Prentice-Hall.

Brusco, S. (1986)
Small firms and industrial districts: the experience of Italy. In: New firms and regional development in Europe, D. Keeble and E. Wever, eds., pp. 184-202. London: Croom Helm.

Buswell, R.J. (1983)
Research and development and regional development: a review. In: Technological change and regional development, A. Gillespie, ed., pp. 9-22. London: Pion.

Coffey, W.J. and Polése, M. (1985)
Local development: conceptual bases and policy implications. Regional Studies 19: 85-93.

Cohen, S.S. and Zysman, J. (1987)
Manufacturing matters: the myth of the post-industrial economy. New York: Basic Books.

Cooke, P. (1985)
Regional innovation policy: problems and strategies in Britain and France. Environment and Planning C: Government and Policy 3: 253-267.

Cooper, A.C. (1985)
The role of Incubator organizations in the founding of growth-oriented firms. Journal of Business Venturing 1: 75-86.

CURDS (1979)
The mobilisation of indigenous potential in the U.K. Report to the Regional Policy Directorate of the European Community. Newcastle-upon-Tyne: Center for Urban and Regional Development Studies.

Dahlman, C. and Westphal, L. (1982)
Technological effort in industrial development - a survey. In: The economics of new technology in developing countries, F. Stewart and J. James, eds., pp. 105-137. London: Frances Pinter.

Dosi, G. (1984)
Technical change and industrial transformation. London: Macmillan.

Dosi, G. (1988)
Sources, procedures, and microeconomic effects of innovation. Journal of Economic Literature 26: 1120-1171.

Dutton, J.M. and Thomas, A. (1985)
Relating technological change and learning by doing. In: Research on technological innovation, management and policy, Vol. 2, R.S. Rosenbloom, ed., pp. 187-224. Greenwich, CT: JAI Press.

Dyson, K., ed. (1988)
Local authorities and new technologies: the European dimension. London: Croom Helm.

Ewers, H.-J. (1986)
Spatial dimensions of technological developments and employment effects. In: Technological change, employment and spatial dynamics, P. Nijkamp, ed., pp. 157-176. Berlin Springer-Verlag.

Ewers, H.-J. and Wettmann, R.W. (1980)
Innovation-oriented regional policy. Regional Studies 14: 161-179.

Florida, R. and M. Kenney (1988)
Venture capital, high technology and regional development. Regional Studies 22: 33-48.

Fosler, F.S., ed. (1988)
The new economic role of American states. New York: Oxford University Press.

Fransman, M. and King, K., eds. (1984)
Technological capability in the third world. New York: St. Martin's Press.

Friedmann, J. (1986)
Regional development in industrialised countries: endogenous or self-reliant? In: Self-reliant development in Europe, M. Bassand et al., eds., pp. 203-216. Aldershot: Gower.

Giaoutzi, M. (1985)
Factors affecting the capacity of technological change: the case of less developed countries. Papers of the Regional Science Association 58: 73-82.

Goddard, J.B. and A.T. Thwaites (1986)
New technology and regional development policy. In: Technological change, employment and spatial dynamics, P. Nijkamp, ed., pp. 91-114. Berlin: Springer.

Gruenstein, J.M.L. (1984)
Targeting high tech in the Delaware Valley. Business Review, Federal Reserve Bank of Philadelphia (May-June): 3-14.

Hükanson, H. (1987)
Industrial technological development: a network approach. London: Croom Helm.

Hall, P. and A.R. Markusen, eds. (1985)
Silicon landscapes. Boston: Allen and Unwin.

Hamfelt, C., and Lindberg, A.K. (1987)
Technological development and the individual's contact network. In: Industrial technological development: a network approach, H. Hükanson, ed., pp. 177-219. London: Croom Helm.

Houttuin, G. (1985)
Venture capitalism. In: Innovation policies, G. Sweeney, ed., pp. 189-196. London: Frances Pinter.

Illeris, S. (1986)
New firm formation in Denmark: the importance of the cultural background. In: New firms and regional development in Europe, D. Keeble and E. Wever, eds., pp. 141- 150. London: Croom Helm.

Johansson, B. (1987)
Information technology and the viability of spatial networks. Papers of the Regional Science Association 61: 51-64.

Johansson, B. and Westin, L. (1987)
Technical change, location, and trade. Papers of the Regional Science Association 62: 13-25.

Jowitt, T. (1988).
Towards silicondale? Lessons from American experience for the Pennine towns and cities. In: Local authorities and new technologies: the European dimension, K. Dyson, ed., pp. 124-139. London: Croom Helm.

Keeble, D. and Kelly, T. (1986)
New firms and high-technology industry in the United Kingdom: the case of computer electronics. In: New firms and regional development in Europe, D. Keeble and E. Wever, eds., pp. 184-202. London: Croom Helm.

Keeble, D. and Wever, E., eds. (1986)
New firms and regional development in Europe. London: Croom Helm.

Kline, S.J. and Rosenberg, N. (1986)
An overview of innovation. In: The positive sum strategy, R. Landau and N. Rosenberg, eds., pp. 275-305. Washington: National Academy Press.

Kok, J.A.A.M. and Pellenbarg, P.H. (1987)
Innovation decision-making in small and medium-sized firms: a behavioural approach concerning firms in the Dutch urban system. In: New technology and regional development, B. van der Knaap and E. Wever, eds., pp. 145-164. London: Croom Helm.

LeHeron, R.B. (1973)
Best-practice technology, technical leadership, and regional economic development. Environment and Planning 5: 735-749.

Lorenzoni, G. and Ornati, O.A. (1988)
Constellations of firms and new ventures. Journal of Business Venturing. 3: 41-57.

Malecki, E.J. (1981)
Science, technology, and regional economic development: review and prospects. Research Policy 10: 312-334.

Malecki, E.J. (1983)
Technology and regional development: a survey. International Regional Science Review, 8: 89-125.

Malecki, E.J. (1987)
The R&D location decision of the firm and "creative" regions. Technovation 6: 205-222.

Malecki, E.J. (1988a)
New firm startups: key to rural growth. Rural Development Perspectives 4(2): 18-23.

Malecki, E.J. (1988b)
Research and development and technology transfer in economic development: the role of regional technological capability. Paper presented at the 1988 European Summer Institute of the Regional Science Association on Theories and Policies of Technological Development at the Local Level, Arco, Italy, July.

Malecki, E.J. (1989)
What about people in high technology? Some research and policy considerations. Growth and Change 20.

Malecki, E.J. and P. Nijkamp (forthcoming)
Technology and regional development: some thoughts on policy. Environment and Planning C: Government and Policy.

Martin, F. (1986)
L'entrepreneurship et la developpement local: une evaluation. Canadian Journal of Regional Science 9: 1-23.

Mason, C.M. (1985)
The geography of "successful" small firms in the United Kingdom. Environment and Planning A 17: 1499-1513.

Meyer-Krahmer, F. (1985)
Innovation behaviour and regional indigenous potential. Regional Studies 19: 523-534.

Miller, H.M., E.E. Brown and T.J. Centner (1986)
Southern Appalachian handicraft industry: implications for regional economic development. Review of Regional Studies 16 (3): 50- 58.

Miller, R. and Coté, M. (1987)
Growing the next silicon valley. Lexington, MA: Lexington Books.

Molle, W. (1983)
Technological change and regional development in Europe. Papers of the Regional Science Association 53: 23-38.

Nelson, R.R. and Winter, S.G. (1982)
An evolutionary theory of economic change. Cambridge, MA: Harvard University Press.

OECD (1985)
Venture capital in information technology. Paris: Organisation for Economic Co-operation and Development.

Oakey, R.P. (1984)
High technology small firms. New York: St. Martin's Press.

Oakey, R.P. (1985)
High-technology industry and agglomeration economies. In: Silicon Landscapes, P. Hall and A. Markusen, eds., pp. 94-117. Boston: Allen and Unwin.

Pavitt, K. (1984)
Patterns of technical change: towards a taxonomy and a theory. Research Policy 13: 343-373.

Pellenbarg, P.H. and J.A.A.M. Kok (1985)
Small and medium-sized firms in the Netherlands' urban and rural regions. Tijdschrift voor Economische en Sociale Geografie 76: 242-252.

Pottier, C. (1985)
The adaptation of regional industrial structures to technical changes.
Papers of the Regional Science Association 58: 59-72.

Pottier, C. (1988)
Local innovation and large firm strategies in Europe. In: High technology industry and innovative environments: the European experience, P. Aydalot and D. Keeble, eds., pp. 99-120. London: Routledge.

Reid, J.N. (1988)
Entrepreneurship as a community development strategy for the rural south. In: The rural south in crisis: challenges for the future, L.J. Beaulieu, ed., pp. 325-343. Boulder, Colorado: Westview Press.

Rogers, E.M. (1982).
Information exchange and technological innovation.
In: The transfer and utilization of technical knowledge, D. Sahal, ed., pp. 105-123.
Lexington, MA: Lexington Books.

Rosenberg, N. (1982).
Inside the black box: technology and economics.
New York: Cambridge University Press.

Russo, M. (1985)
Technical change and the industrial district: the role of interfirm relations in the growth and transformation of ceramic tile production in Italy.
Research Policy 14: 329-343.

Sabel, C., Herrigel, G., Kazis, R. and Deeg, R. (1987)
How to keep mature industries innovative. Technology Review 90, 3: 27-35.

Schell, D.W. (1983)
Entrepreneurial activity: a comparison of three North Carolina communities. In: Frontiers of entrepreneurship research 1983, J.A. Hornaday et al., eds. pp. 495-518.
Wellesley, MA: Babson College Center for Entrepreneurial Studies.

Schell, D.W. (1984)
The development of the venture capital industry In North Carolina: a new approach.
In: Frontiers of entrepreneurship research 1984, J.A. Hornaday et al., eds., pp. 55-72.
Wellesley, MA: Babson College, Center for Entrepreneurial Studies.

Segal, A., ed. (1987)
Learning by doing: science and technology in the developing world.
Boulder, CO: Westview Press.

Senker, J. (1985)
Small high technology firms: some regional implications.
Technovation 3: 243-262.

Shapero, A. (1971)
An action program for entrepreneurship. Report prepared for the Ozarks Regional Commission. Austin, Texas: Multi-Disciplinary Research Inc.

Shapero, A. (1984)
The entrepreneurial event. In: The environment for entrepreneurship, C.A. Kent, ed., pp. 21-40. Lexington, MA: Lexington Books.

Shapero, A. (1985)
Managing professional people. New York: The Free Press.

Smilor, R.W., G. Kozmetsky, and D.V. Gibson, eds.(1988)
Creating the technopolis. Cambridge, Mass.: Ballinger.

Spence, A.M. and H.A. Hazard, eds. (1988)
International competitiveness. Cambridge, Mass.: Ballinger.

Stöhr, W.B. (1986)
Regional innovation complexes.
Papers of the Regional Science Association 59: 29-44.

Sweeney, G.P. (1985)
Innovation is entrepreneur-led.
In: Innovation policies: an international perspective, Sweeney, ed., pp. 80-113. London: Frances Pinter.

Sweeney, G.P. (1987)
Innovation, entrepreneurs and regional development.
New York: St. Martin's Press.

Tatsuno, S. (1986)
The technopolis strategy. New York: Prentice Hall Press.

Teece, D.J. (1986)
Profiting from technological innovation: implications for integration, collaboration, licensing and public policy. Research Policy 15: 285-305.

Thomas, M.D. (1987)
The innovation factor in the process of microeconomic industrial change: conceptual explorations. In: New technology and regional development, B. van der Knaap and E. Wever, eds., pp. 21-44. London: Croom Helm.

Thomas, M.D. (1988)
Technical entrepreneurship in high technology industries:
exploratory conceptualizations of firm formation and location processes.
Paper presented at the North American Meetings of the Regional Science Association, Toronto, November 1988.

Thwaites, A.T. (1982)
Some evidence of regional variations in the introduction and diffusion of industrial products and processes within British manufacturing industry.
Regional Studies, 16: 371-381.

Varaiya, P. and Wiseman, M. (1981)
Investment and employment in manufacturing in U.S. metropolitan areas 1960-1976.
Regional Science and Urban Economics 11: 431-469.

von Hippel, E. (1988)
The sources of innovation. New York: Oxford University Press.

Watts, H.D. (1981)
The branch plant economy. London: Longman.

Wetzel, W.F. (1986)
Informal risk capital: knowns and unknowns. In: The art and science of entrepreneurship, D.L. Sexton and R.W. Smilor, eds., pp. 85-108.
Cambridge, MA: Ballinger.

Wever, E. (1986)
New firm formation in the Netherlands. In: New firms and regional development in Europe, D. Keeble and E. Wever, eds., pp. 54-74. London: Croom Helm.

3.4 Industrial Innovation and the Spatial Division of Labour

Franz-Josef Bade

1 The Innovative Imperative

Frequently, it is postulated that innovation is an essential and undisdisputable precondition for the competitiveness of firms. Usually, this kind of "innovative imperative" is founded on two interwoven lines of argument. On the one line the increasing pressure of international competition is stressed. The other line of argumentation concerns the speed of technical progress and its consequences for the development of new products and production processes. It is a quite popular opinion among business people that the life cycles of products are continuously becoming shorter due to the accelerating rate of technical innovation. Or, as one chairman of Dupont once formulated: "Lead time is gone. There is no company so outstanding technically today that it can expect a long lead time in a new discovery." (see Fig.1).

The conclusion which is drawn from the increasing pressure of competition and the shortening of life cycles is the growing danger of missing the market changes. As life cycles shorten, the possibilities for gaining enough of a market share in order to reach the break-even point get smaller. Thus, the innovative imperative appears to be the only adequate answer: There is no other way to survive than being first with new technological developments.

However, this imperative is not uncontroversial. There is a long discussion about whether the strategy of being the "first on the market" is necessarily the most successful in economic terms. It cannot be excluded that the monopolist earnings due to the newness of products do not cover the additional costs of research and development, for, by definition, the costs of research and development are rather insecure and hard to be estimated.

Figure 1: Life cycles of products (Changes 1979-1984)

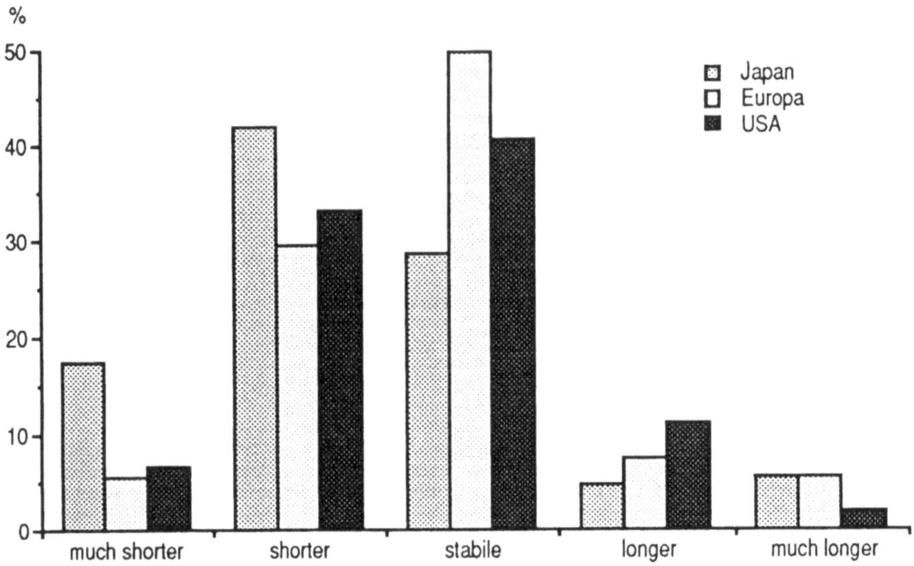

In fact, the empirical evidence is not unequivocal. Of course, we can observe cases where the strategy of "being first" was successful. But there are also a lot of cases where a more temporizing strategy was successful. One example is the strategy of the "fast second" who imitates the new product, but uses the time lag for improving the product quality. Therefore, it is not surprising that in statistical investigations, the relationship between innovation and economic performance has not been proven unequivocally. It is true that most of the firms with a high rate of innovation have a better economic performance than the average. But the reverse relationship does not hold: there are a lot of firms with a lower innovation rate which, nevertheless, show an above-average growth.

Beyond these general doubts about the strategy of "being first", there are conceptual reasons for the ambivalent results of statistical analysis. First, in many investigations the idea of innovation is restricted to the output of the innovation process. Thus arises the principle difficulty that for each firm in the analysis an estimation should be made about the innovativeness of the complete product programme; the researchers hardly have the necessary knowledge to undertake this task. The solution is either to leave the estimation to the manager (e.g., "how large is the portion of new products?"), or to concentrate on a few products which are regarded as innovative by some (expert) standards. In both cases, the relationship between innovation and economic success remains obscure: In the first case, it can hardly be guaranteed that all respondents have the same understanding of the newness of products. In the latter, the economic importance of the remaining (not analysed) products, for example, their impact on profit, is necessarily outside of the analysis (see Fig.2).

Second, the selection of goods or processes regarded as innovation is frequently based on rather technical standards. However, technical developments are only one part of the changes in the environment of a firm. Other parts, above all, changes in market areas such as sales, purchases, finances or labour, are at least of the same importance for successful survival. Likewise, it is generally accepted in studies that one essential precondition for economic success is a deliberate product/market concept. The primary character-

Figure 2: Components of environment

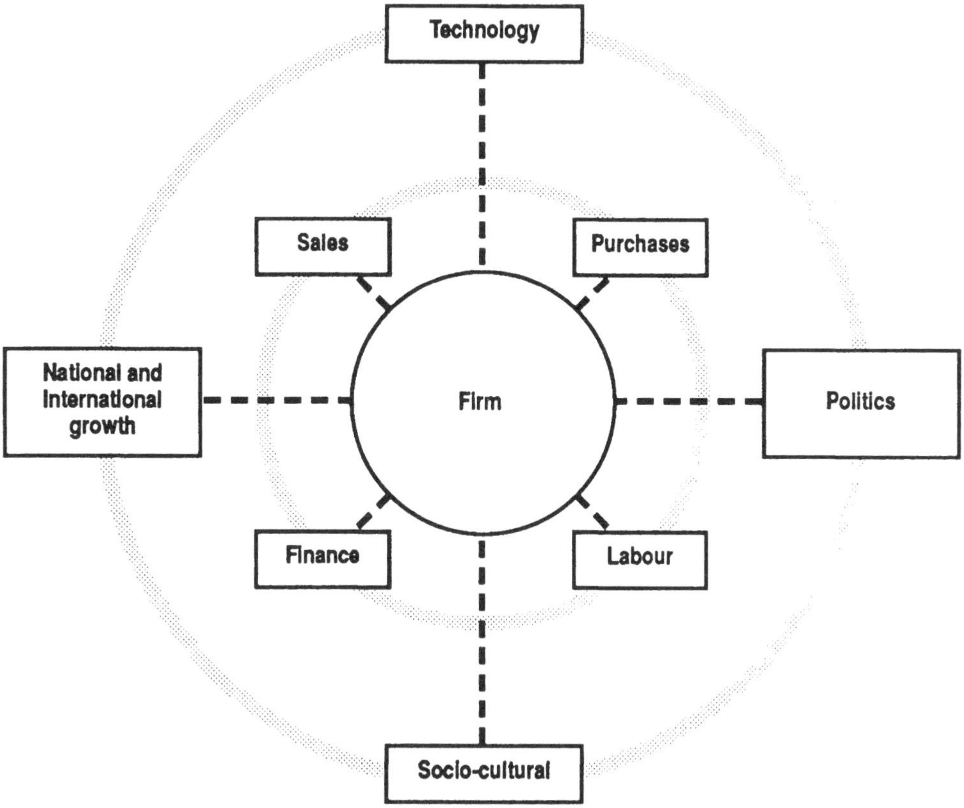

istic, however, of such a market strategy is not its technology orientation. Instead, a product/market concept tries to integrate the requirements of the different market areas and shows a way to make the best profit out of the specific condition.

2 Producer Services as Input Activities for Innovation

Many of the above mentioned problems of strategy and analysis can be avoided if the narrow perspective on innovation as result or output is given up. Considering the final economic impact, a wider interpretation of innovation seems to be more productive. Such a wider context focuses the interest on the input side, on the capacity of firms to adopt to the changes of environment in an active innovative way. Thus, the primary aim of the wider understanding is not to judge whether a product is an innovation or which strategy is more successful. On the contrary, the perspective is directed to those activities of a firm which may enforce its capacity to act in an innovative way whatever this way is. Consequently, innovations, narrowly and rather technically defined, are of minor, or better, of indirect importance; they are only one of several more or less adequate possibilities of acting innovatively.

Obviously, there is no simple answer to the question of what are the determinants of the innovative capacity of a firm. However, among the activities which may be relevant for the innovative capacity of firms are, above all, those firm-internal services which we call producer services such as marketing, consulting, research and development, and other comparable activities characterized by a long-term perspective. Of course, one has to be aware that the context in which innovations originate is much too complex to be reduced to one or two singular factors of influence. Nevertheless, such services should be regarded as essentials for innovative as well successful actions.

Meanwhile, there is some empirical evidence which underlines the supposed importance of such producer services. Besides the rather obvious relationship between research and development activities and (narrowly defined) innovations, in some studies, the necessity of planning and marketing activities is stressed. Finally, however, I think it impossible and even unnecessary to define such innovation-relevant activities in a detailed, and at the same time, universal description.

The primary feature of such activities is "looking ahead" characterized by an "offensive" search for possibly relevant information. But there is no universal answer regarding where this looking ahead is necessary and which particular activities are to be untertaken. Without the very detailed knowledge of the demand and supply conditions of a firm, for example, a recommendation of particular services is of little use in most cases. There is, to a certain extent an analogy here with the discussion about market strategies. Without detailed market information it is worthless to discuss the appropiateness of a firm's market strategy. Primarily, seen as a necessary minimal condition, it is important whether the firm has a deliberate product/market concept. Likewise, it is the existence of (some) innovation-relevant activities that is of primary importance for the analysis of the innovation capacity of firms. It is true that, on the basis of their longer-range information character, it is possible to indicate a group of such activities in rough outlines. But the importance of these activities may differ from firm to firm. In technology-intensive production, for example, in non-ferrous metal industries or in the chemical industry, research and development have obviously a greater influence. In industries with rather standardized products, on the other hand, competitive advantages are mainly created by product differentiation due to some extra product features. Therefore, in industries like clothing or, nowadays, the watch industry, marketing activities seem to be of much greater relevance.

3 Empirical Evidence on the Economic Influence of Producer Services

Thus, the exact delineation of innovation-relevant activities is inevitably to some extent voluntary. That is why in our own empirical analysis, we have tried to avoid defining these possibly innovative activities too narrowly. Instead, we have constructed a group of activities which, being regarded over all economic sectors, appear to be rather heterogeneous; the group reaches from research and development over strategical planning and up to marketing and EDP.

We can observe a strong relationship between the intensity of such producer services and economic development. Sectorally considered, apart from some industries with rather state-regulated markets (i.e., aviation or steel), those industrial groups with a higher share of producer services have a clearly better performance, measured by their development of value added, than industries where relatively fewer employees are involved in such possibly innovation-relevant activities (Fig.3). The same relationship holds within a regional perspective. Agglomeration centres with an above-averge employment turnover are generally characterized by a higher intensity of producer services (Fig.4).

In some discussions, it is argued that the higher intensity of producer services observed in Munich or the lower share in the Ruhr region, is mainly a result of the regional differences in sectoral structure. It is true that, for example, Munich has a larger share of in-

Figure 3: Influence of producer services on the development of industrial value added 1976-1986

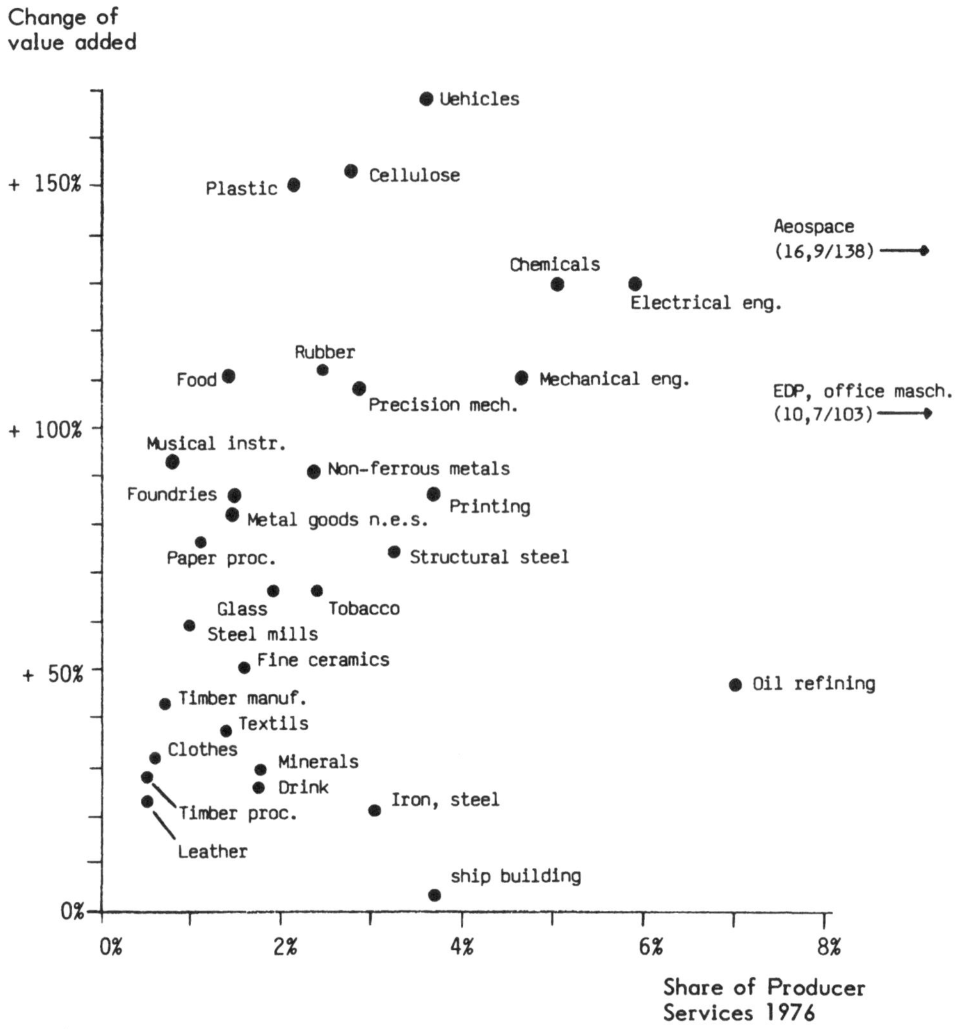

Legend: Value added: Change of value added in actual prices 1976 – 1986
Share of producer services: Research and development, consulting, marketing and EDP as share of total employment in the respective industry 1976

dustrial groups with a higher intensity of innovation-relevant activities than the Ruhr region. But in Fig. 5, it can be observed that the relationship between regional employment change and regional intensity of producer services even holds when it is differentiated by industrial groups.

Figure 4: Influence of producer services in the agglomeration centres 1976-1986

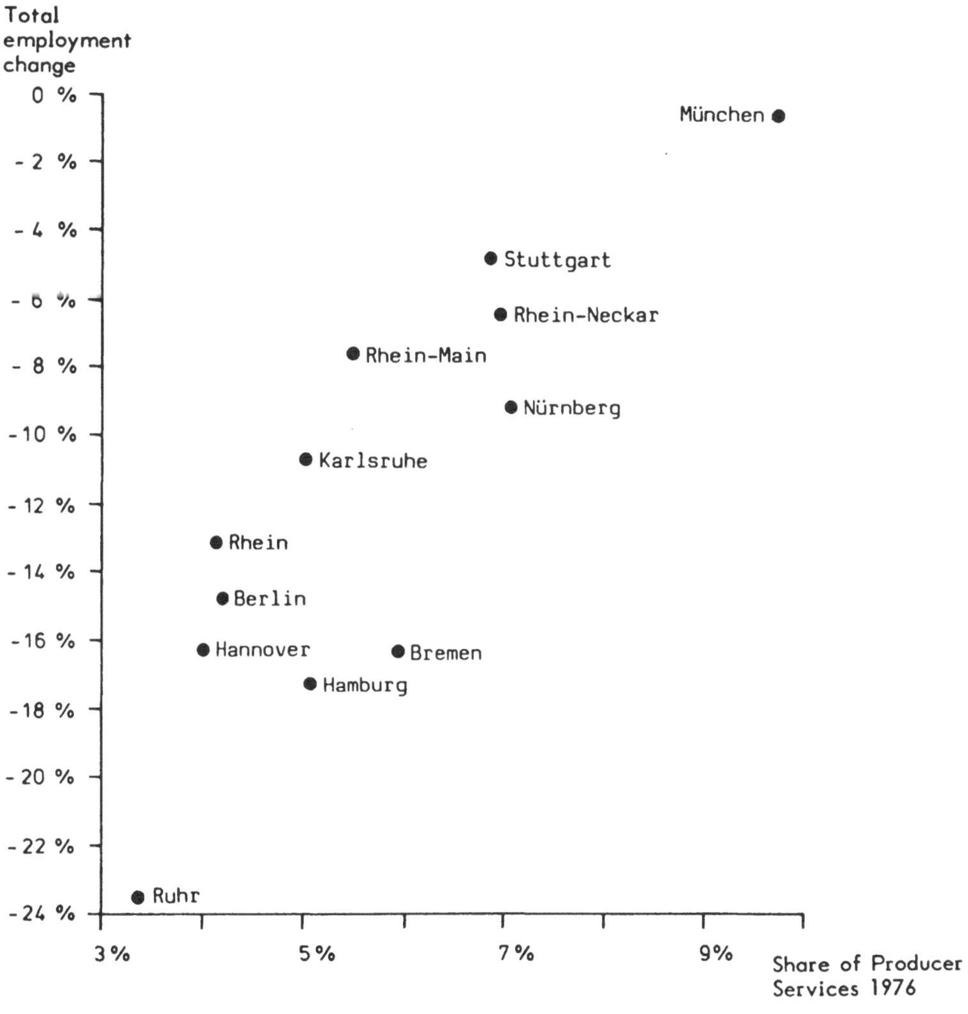

Legend: Total employment change: Change of total employment in manufacturing industries between 1976-1986

Share of producer services: Research and development, consulting, marketing and EDP as share of total employment in the respective industry 1976

4 Regional Variations in the Influence of Producer Services

Given the empirical evidence of producer services, the future of regions outside the large metropolitan areas appears to be rather sobering. For, there is a clear-cut spatial division of labour between metropolitan centres and the other parts of the Federal Republic of

Figure 5: Influence of producer services on selected industries in the agglomeration centres 1976-1986

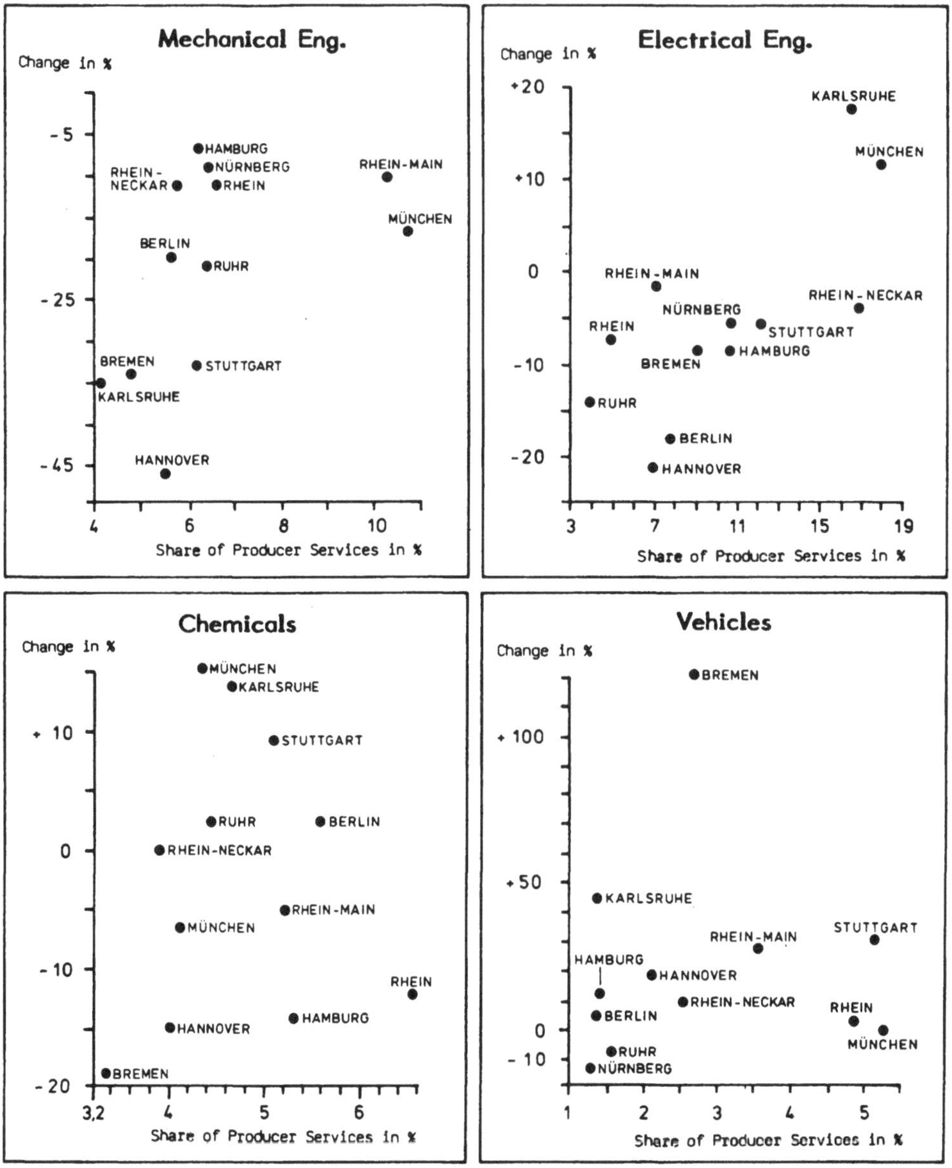

Legend: Total employment change: Change of total employment in the respective industry between 1976-1986

Share of producer services: Research and development, consulting, marketing and EDP as share of total employment in the respective industry 1976

Figure 6: The spatial division of labour in the FRG
Share of respective total regional employment, expressed as deviation from the respective average of the FRG

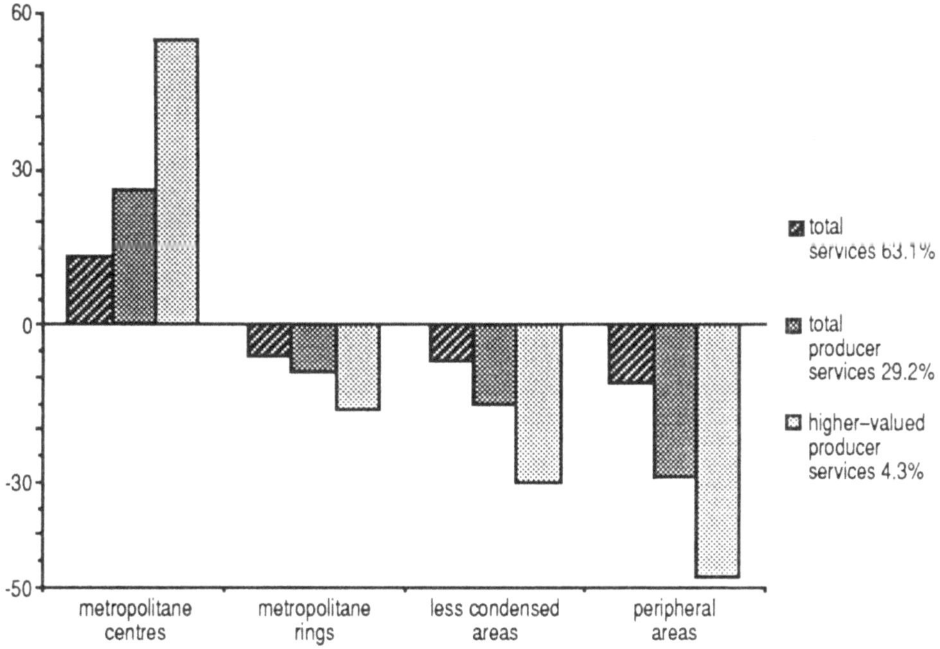

Germany (FRG): The density of producer services such as research and development is three times as high in the metropolitan centres as in the peripheral regions (Fig. 6). On the other hand, however, the FRG like other industrialized countries is characterized by a strong decentralization process in the spatial distribution of economic activities. Among all types of areas, metropolitan centres are the main loser in the change of spatial structure. On the contrary, peripheral regions as well the rings of metropolitan areas have had the highest gains in employment (Fig. 7).

Thus, there seems to be an apparent contradiction between the performance of regions and the supposed influence of producer services. Actually, this contradiction dissolves when the different types of areas are considered together. From Figs. 8 and 9 it can be seen that in all types of areas (Fig.8: peripheral areas and metropolitan centres; Fig.9: metropolitan rings and centres) the positive relationship between density of producer services and employment change holds. But there is an obvious large difference in the degree of influence: the effect of (additional) producer services in the peripheral areas is much higher than in the metropolitan areas.

Figure 7: Changes in the spatial structure of the FRG 1976 – 1986

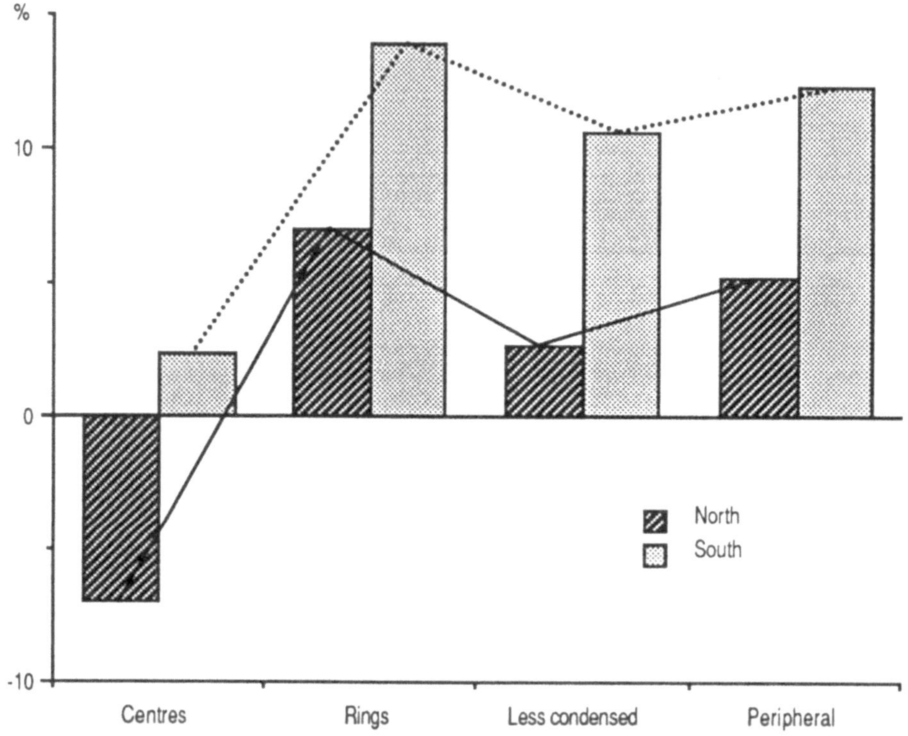

Figure 8: Regional disparities in the influence of producer services 1976-1986
Agglomeration centres and rings

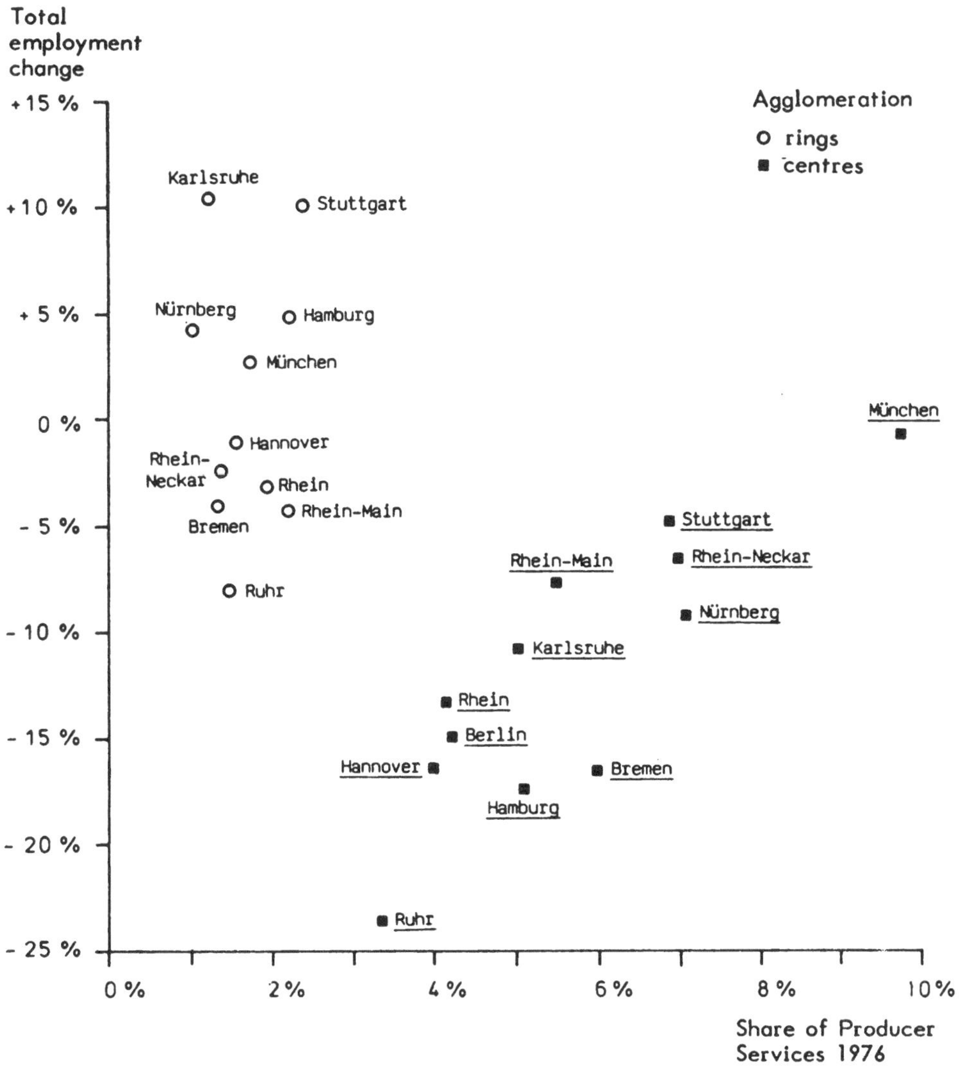

Legend: Total employment change: Change of total employment in manufacturing industries between 1976-1986

Share of producer services: Research and development, consulting, marketing and EDP as share of total employment in the respective industry 1976

Figure 9: Regional disparities in the influence of producer services 1976-1986
Agglomeration centres and peripheral areas

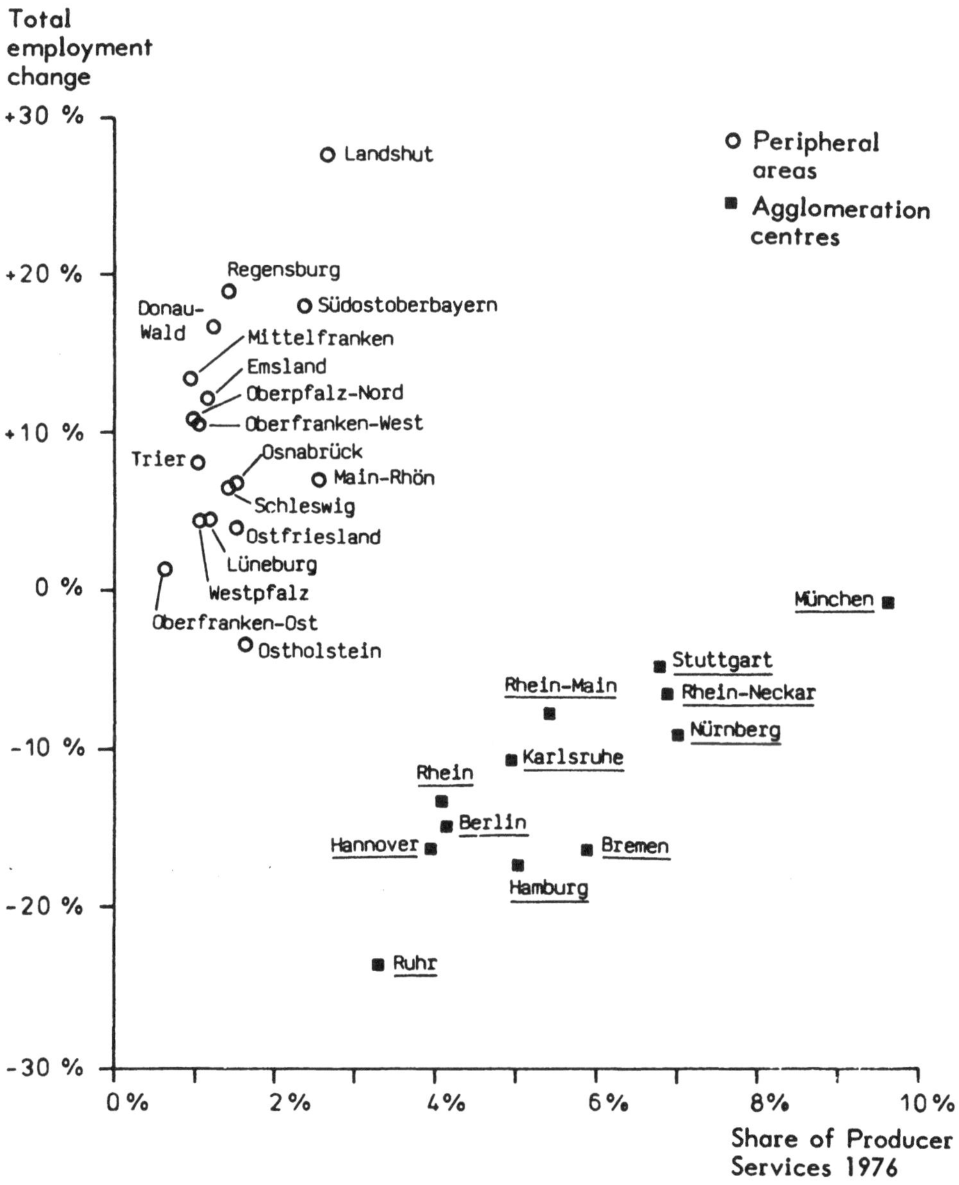

Legend: Total employment change: Change of total employment in manufacturing industries between 1976-1986

Share of producer services: Research and development, consulting, marketing and EDP as share of total employment in the respective industry 1976

3.5 Endogenous Development between Myth and Reality: Pre-Requisites for Endogenous Development Strategies

Ernst A. Brugger

In scientific and political circles there is increased talk of endogenous development and of the strategies involved in implementing it. Everyone is in favour of it, no one against, which seems suspicious. Does the concept really encompass so much meaning that the whole world comes under its spell? Or is it rather the other way round, where everbody can wrap up his or her particular aims and ideas so that "endogenous development" comes to represent nothing more than an empty shell? The concept floats somewhere between reality and myth, perhaps in the end it is rather a realistic utopia with its specific form and colour depending on the particular initial interests and ideologies.

1 Theses

The following three theses are to be seen against the background of growing worldwide economic, cultural, and political integration. Even in the sphere of ecology the picture of "spaceship earth" becomes more and more obvious. Progress in communication and technology is leading to the rapid development of worldwide networks. Finance and know-how (to name just two agents of production) become increasingly ubiquitous, at least as potential factors of economic development. Growing internationalization in nearly all areas means an increasingly functionalist approach. As a sort of counter-movement, local and regional territorial orientations, in the sense of regionalism movements or of movements back to small networks are also gaining ground. The trade-off between a functional and a territorial orientation is receiving growing consideration.

Against this background the aims of endogenous regional development are to be defined precisely within a concept of territoriality: a region, however delimited it may be, determines its own economic, political, cultural, and even ecological development no matter how this development may be defined. In the end, it is this fairly high degree of self-determination of a region that defines the goal setting and enforcement of the strategies of endogenous development.

1.1 The First Thesis

There are basically two different ways to reach the target. The first one is an "outward-oriented endogenous development strategy", and currently this is the approach most under discussion. A region has or develops a potential for something which is of interest to the rest of the world, but which it will not let out of its control. The goal, after all, is an offensive policy regarding relations with the rest of the world, a policy to be shaped for the region and to the region's greatest advantage. The second possible way is an "inward-oriented endogenous development strategy" where a region, usually a poorer and peripheral one, sets its development goals at such a low level that it is of no interest to the rest of the world, nor does it want to be. This type of modesty is generally very implicit in its concept inclusive of a marked ecological orientation. This type of goal setting leads to a sort of decoupling, to a development with practically no growth, a sort of qualitative zero growth.

1.2 The Second Thesis

Despite their different orientations both possible ways presented here as ideal models require three equally necessary, although not sufficient, preconditions:

1. Control of the interregional exchange of goods, services, production factors and ideas. This requires a relevant "bargaining power".
2. A clear development-policy is a necessary basis for this bargaining power. A region has to know what it wants and has to have the capacity to make decisions.
3. A high degree of innovation: this means economic innovation, but also innovation in organizational, planning and political innnovation.

1.3 The Third Thesis

The three preconditions, i.e. bargaining power, decision making and innovation capacity, are hardly ever or very rarely, fulfilled. The realization of endogenous development strategies is therefore minimal. Nevertheless, the goal setting may, under certain conditions take a clear orientation in the development process and the region's development policy.

In the following part the three conditions mentioned are to be more closely examined.

2 Bargaining Power

A region with an outward-oriented endogenous development strategy opens up to world-wide competition and wants to strengthen its position there accordingly. Therefore, it will try to carry through a strategy of competitiveness, especially in the economic sphere. It will also orient its basic framework and its promotional instruments towards fostering innovation-intensive, technically advanced and, if possible, internationally active businesses. Coupled with the great potential of those firms, the prerequisite for success also includes a large, dense and flexible labour market, a liberal economic policy and a very tight, active network in the economic sphere as well as between it and the political sphere.

In fact, only the world's largest agglomerations show these characteristics. Only they have the size, density and quality, to take and hold a world market position in the sense described above, at least in the economic field. Smaller agglomerations have a chance to succeed only if they systematically link themselves to one of the 10 to 12 largest world agglomerations and if they involve themselves in a close and ever closer network and input-output relationship. In contrast, with an inward-oriented endogenous development strategy, we proceed on the assumption that a region with a small development potential, e.g. with a rather rural and peripheral character, redefines the quality of its development goals. Ultimately, and compared with the rest of the world these are alternative goals. Neither economic growth nor competitiveness stand in the foreground, but an holistic outlook on life with strong cultural, social and increasingly ecological criteria. This strategy means a sort of normative disintegration the so-called "elegant poverty concept" where bargaining power is weak, compared with the outward-oriented strategy, but can be weak since there is not much to bargain for anyway. Nevertheless, in the tourist industry for instance, it has to be strong enough to avoid too much outside interference.

3 Decision-Making

As a rule, this second precondition is extremely difficult to achieve in an outward-oriented endogenous strategy. Political structures and decision-making in the world's largest agglomerations are characterized by a high degree of plurality and heterogeneity such as city core vs. hinterland problems, and dichotomies of economic and socio-cultural orientation. Precisely the economically well-endowed regions probably show the least political homogeneity which, after all, is the precondition for a high capacity for political decison making in democratic structures. However, it is possible that the concept of political decision making changes in these areas upon which the dynamic economic development enforces political decisions. Objective constraints, bottlenecks, etc. play an increasingly dominant role. The political decision making to promote a dynamic and competitive economy can of course de facto also be achieved in this way.

In contrast, political decision making with an inward-oriented strategy ultimately means renouncing the normal value-added process. This can only succeed under certain preconditions as the result of a democratically made process and decision. Above all, democratic structures must function in such regions for some time, e.g. where there is a combination of superannuated local labour and young alternative newcomers, till basically a common fundamental attitude consolidates to a degree where it becomes oriented towards endogenous development with its less ambitious economic goals. Yet a common orientation is not enough, the object being the implementation of goals, in other words, the relevant organizational capacity has to be developed.

4 Innovation Capacity

Within an outward-oriented endogenous development strategy, the economic innovation capacity is the motor for the aspired international competitiveness. The concepts of innovative development strategies are sufficently known. Nevertheless, three relatively new directions are emerging from current discussion and will also be of significance for the shaping of regional policy instruments: the importance of stimulating productive services; the significance of the networks between such services, the link of finance and the entrepreneurial network; the increasing significance of the politico-economic and general enabling conditions of a region, the efficiency of its administrative and infrastructural conditions (e.g. in communication). These three key concepts also signify that the direct support of enterprises becomes less important with this strategy, compared with the more basic concepts in the field of network development including the enabling conditions for the arrangement of such networks.

However, a high capacity for innovation is also needed by an inward-oriented endogenous development strategy. Economically an inward orientation can only work when the local and intraregional networks, markets and consequently the value-added mechanisms are being used and cared for. Culturally this strategy can only be durably implemented with a strong local regional identity and appropriate dynamics for cultural activities.

New concepts - such as an internalizing mechanism for the consumption of ecological goods - will have to serve as guidance in the ecological field. The entire regional development policy will have to pay great attention to the linkage between the ecological and economic spheres. Only few practical examples are available with regard to most of the mentioned key words. Thus, innovations for the implementation of this strategy particularly affect organizational structures, ways of inter regional co-operation, etc. John Friedman calls this approach "radical social practice", for which only individual experiences and some groups' programmes are at hand. Indeed, as it stands, it is hardly imaginable how a region could enforce such a development strategy in the long term.

5 Conclusion

The short discussion on the three preconditions for implementing either of the two possible models of endogenous development strategy shows that basically the three theses are correct.

After all the concept of endogenous regional development can be of great significance for the approach of two completely different patterns. However, as the analysis has tried to show, this normative clarity and its ensuing conceptual precision are neccesary preconditions for even a slight chance of realization of either model.

Finally, the question remains whether the two possible models of endogenous development strategies exclude one another within given countries. Here also two possible patterns can be described:

The regions of a state, within the limits of their decision-making and bargaining powers, opt for one of the two strategies described above or perhaps for a combination of both. This would lead to a picture of internationally oriented economic regions, on the one hand, and unlinked, inward-oriented regions on the other. Between the two types of regions a sort of complementary relationship would develop with growing significance in the ecological sphere: inward-oriented regions would increasingly become green compensatiory areas for outward-oriented regions. This scenario implies a deep functional complementarity between the regions.

However, the following picture is more likely to emerge:

In most European regions both orientations will appear among different social and political groups. This means that ultimately the discussion would be less concerned with homogeneous, sharp outlined regional strategies than with group-specific goals and strategies within the same territorial units. Regional development policy would then develop and simultanously carry into effect very different group-specific concepts although objectives and strategic orientation may be of contrasting nature. Different social groups would then select one of the strategies for themselves while regions would develop into a sort of service unit for completely different needs. In that case of course the above concept of endogenous development would not occur, at least not by its own definition.

Empirical evidence is available for both models described. It will be very exciting to watch which model will be used in which circumstances and which endogenous development strategy will in fact be implemented. Reality alone will show the utopian potentials and risks which make the concept of endogenous development so fascinating.

4 Instruments of Innovation Oriented Regional Policies

4.1 Financial Instruments for the Promotion of Regional Development

Kevin Allen and Douglas Yuill

1 Introduction

This paper is concerned with identifying key recent trends in the development of financial instruments for resolving regional problems in the European Community countries. The paper is in three sections. In section 2 four key trends are identified in respect of the main regional incentives on offer in the Community countries. The forces explaining these trends are discussed in section 3. Finally, conclusions are drawn in section 4, particularly about how the trends identified impact on innovative high-tech projects and firms. This final section also considers how regional incentive instruments might best be designed to increase their leverage on such projects.

2 Key Regional Incentive Trends

Since the late 1970s we have been involved in the detailed monitoring of regional incentive policy in the European Community countries. This work centres on the production of our annual report, European Regional Incentives and is funded by government departments and agencies in six European countries - France, Germany, Italy, the Netherlands, Sweden and the United Kingdom. This year saw the production of our ninth report. Given such a timespan, it is instructive to look back over the past decade with a view to identifying some of the key regional incentive trends which have evolved. These are many and varied. However, in the context of this volume, there are four main developments which are worth highlighting.

First, and perhaps the key trend as far as regional incentive policy is concerned, there has been a very substantial reduction in the importance of automatic regional incentive instruments (with fixed rates of award and overt conditions of award) and a major growth in the significance of discretionary schemes (where the rate of award is at the discretion of the awarding body and is up to a set maximum). This trend is documented in broad terms in Table 1.

Table 1: Degree of administrative discretion in regional incentive packages in 1980 and 1988

Country	Mainly discretion	Automatic/discretion mix	Mainly automatic
1980 Position			
Belgium	•		
Denmark	•		
France		•	
Germany		•	
Ireland		•	
Italy			•
Luxembourg	•		
Netherlands		•	
UK		•	
1988 Position			
Belgium		•	
Denmark	•		
France	•		
Germany	•		
Ireland	•		
Italy			•
Luxembourg	•		
Netherlands		•	
UK	•		

Source: The above table summarises the position described in Yuill, D. and Allen, K. (eds), European Regional Incentives 1980, Table 11; and Yuill, D. Allen, K., Bachtler, J. and Wishlade, F. (eds), European Regional Incentives: 1989 edition, Table 4.

Table 1 shows that, of the nine Community countries in 1980, only one-third had regional incentive packages which were basically discretionary in administration (Belgium, Denmark and Luxembourg). The remaining countries generally had a very significant automatic component to their regional incentive packages, normally in the form of an automatic capital grant providing a predictable base to the package. In contrast, two-thirds of the same nine countries now have regional incentive packages without any noteable automatic component. The pendulum has swung firmly (and far) towards regional incentive systems which are, for the most part, discretionary in their administration.

How has this happened? The main developments since 1980 are chronicled in Table 2. From Table 2, it can be seen that, over the period considered, no fewer than four countries have abolished key automatic regional incentives - Export Sales Relief in Ireland, the Investment Allowance in Germany, the WIR Regional Allowance in the Netherlands and the Regional Development Grant in Britain (and its Northern Ireland equivalent, the Standard Capital Grant). In addition, the Regional Policy Grant in France has been trans-

Table 2: Degree of administrative discretion in regional incentive packages: key changes between 1980 and 1988

Country	Change
Belgium	In the course of 1987, reviews were made of both the Flemish and Wallonian regional incentive packages. While discretionary in theory they had become semi-automatic in practice. They were revamped to strengthen the predictability of basic awards while retaining discretion for projects of particular importance.
Denmark	No major change.
France	New regional incentive legislation introduced in July 1987 shifted the Regional Policy Grant (PAT) scheme away from being a semi-automatic investment grant towards a discretionary incentive aimed at influencing key location decisions.
Germany	The automatic base component of the German regional incentive package, the Investment Allowance, has been abolished in respect of projects started after 31 December 1988.
Ireland	Export Sales Relief, an automatic fiscal concession and key component of the Irish incentive package of the time, was withdrawn at the start of 1981.
Italy	No major change, although it is worth noting that, following the 1986 Mezzogiorno Law, so-called planned bargaining in respect of important innovative projects has the previously wholly automatic Mezzogiorno package.
Luxembourg	No major change.
Netherlands	The WIR Regional Allowance, an automatic fiscal concession, was abolished as from September 1983.
United Kingdom	As from 31 March 1988 the automatic base component of the regional incentive package was abolished (Regional Development Grant in Britain, Standard Capital Grant in Northern Ireland). Although the new Regional Enterprise Grants are basically automatic in administration they are very small-scale and are restricted to Development Area firms with fewer than 25 employees.

Source: Yuill, D. and Allen, K. (eds), European Regional Incentives, various editions (1980-1989).

formed from a semi-automatic investment grant to a discretionary incentive aimed at influencing key location decisions. All told, the move away from automatic regional incentives as the base component of regional incentive packages has been very wide-ranging. Indeed, only in Italy does automatic regional assistance continue to play a central role in the regional incentive system.

Despite the strong trend there has been in the course of the 1980s towards more discretionary regional incentive schemes, it would be misleading to suggest that developments have all been in the one direction. For instance, Table 2 makes it clear that the recent (1987) regional incentive reviews in both Flanders and Wallonia have resulted in a significant reduction of administrative discretion in Belgium.

In Wallonia, dissatisfaction with a system which, while discretionary in theory, had become semi-automatic in practice, led to the establishment of a two-tier incentive system which distinguishes between predictable "automatic" assistance (available to the majority of applicants) and discretionary contractual assistance (focusing on some 15 projects a year considered to be of strategic importance because of their size and/or type of investment). In a similar vein, the revised Flemish incentive system aims to make rates of award far more visible (being expressed as a grant-equivalent percentage of gross eligible investment) and also much more predictable.

Nor have such developments been restricted to Belgium. Although not covered in Tables 1 and 2, the new regional incentive packages in both Spain and Portugal (following their entry into the European Community) are very much more grant-oriented than the schemes they replaced. Far more so than was previously the case, rates of award in these new Community countries are visible and (within certain specified ranges) predictable.

The arguments for introducing more discretion into regional incentive administration seem clearcut. First and foremost there is the budgetary saving associated with the move away from automatic assistance. In addition, increased discretion reflects a desire to make the incentives on offer more cost-effective. In particular, discretionary administration aims to avoid the windfall gains often associated with automatic schemes. However, there are also certain worries about discretionary assistance, not least a concern about whether such schemes can be administered in a truly discretionary manner. Moreover, the positive features of automatic assistance such as visibility, predictability and neutrality between different firm sizes, are not easily replicated within a discretionary scheme which, if it is to operate at the margin, must - prior to the award decision being made leave recipients unclear about just how much, if anything, they are likely to receive. Recent developments in a number of Community countries, and noted above, suggest that the policy pendulum mentioned earlier may have swung too far towards discretionary assistance and that the need for regional incentives to be visible and, above all, predictable in award, may be given increasing weight in the years to come.

So much for the trends towards more discretion in the administration of regional incentive policy. A second important development of the 1980s has seen a growing degree of selectivity and rate discrimination in the regional incentives on offer. This is illustrated by Table 3 which shows the extent of explicit rate discrimination in the main regional capital grants in the Community countries in 1980 and 1988. Table 3 shows a particular

Table 3: Explicit rate discrimination in main regional grants in 1980 and 1988

	Explicit rate discrimination by:				
Country	Area	Sector	Size	Project Type	Employment
1980 Position					
Belgium	•	•			
Denmark					
France	•			•	
Germany	•				
Ireland	•				
Italy	•	•	•		
Luxembourg	•				
Netherlands	•		•		
UK	•				
1988 Position					
Belgium	•	•			•
Denmark	•				
France	•				
Germany	•			•	
Ireland	•			•	
Italy	•	•	•		
Luxembourg	•				
Netherlands			•	•	
UK	•				

Source: The above table summarises the position as described in detail in Yuill, D. and Allen, K. (editors), European Regional Incentives 1980 and Yuill, D., Allen, K., Bachtler, J. and Wishlade, F. (editors), European Regional Incentives: 1989 edition. The main changes which have taken place are highlighted in Table 4.

growth in spatial rate discrimination and also in the degree of rate discrimination by project type.

The developments which underpin the trends highlighted in Table 3 are set out in Table 4, which shows clearly that, in the course of the eighties, there was a significant increase in spatial rate discrimination most obviously in Luxembourg, Italy and Denmark. More recently, there has also been a major growth in rate discrimination by project type with the setting up of projects now being favoured over extensions of Germany, Ireland and The Netherlands. In Belgium, too, the 1987 regional incentive reviews resulted in a notable increase in rate discrimination by location, sector, employment and type of investment. Viewing the Community countries as a whole, there is now significantly more explicit rate discrimination than was the case a decade ago.

Of course, as with all such generalisations, there are exceptions. One such is The Netherlands where rate discrimination between different locations has recently been eliminated the direct consequence of the major cutback there has been in the coverage of the designated Dutch problem regions. Another exception is France where the level of explicit rate discrimination has fallen compared to the position at the start of the decade, but where the greatly increased administrative discretion since 1987 (see Table 2) has resulted in an even more targetted policy than was previously the case. The French Regional Policy Grant is now very much focused on important large-scale projects and relocations away from the Paris region.

The French case is interesting because it highlights the narrowness of the picture given by Tables 3 and 4, with their emphasis on explicit rate discrimination. The tables take no account of the fact that, as already mentioned, there has been a very major increase in the level of administrative discretion in the course of the 1980s. Like explicit rate discrimination, an important feature of administrative discretion is that it tends to be used to direct spending towards what are considered to be key policy targets. The growth of administrative discretion and explicit rate discrimination in tandem has clearly added considerably to the overall level of selectivity in incentive award. Far more so than was the case at the start of the decade, regional incentives are directed towards specific sectors, locations and project types, with innovative projects being particularly favoured.

The two regional incentive trends of the 1980s considered so far have focused primarily on the administration of the available regional incentives. A third important trend relates to the type of incentive on offer. Whereas at the start of the decade there was a very heavy reliance on investment-related incentives (in particular, capital grants, soft loans

Table 4: Explicit rate discrimination in main regional grants: key changes between 1980 an 1988

Country	Change
Belgium	While the basic regional incentive legislation has remained unchanged over the period (the Economic Expansion Act 1970), policy reviews in the course of 1987 resulted in more explicit rate discrimination in practice in both Flanders and Wallonia. In Flanders, a basic award distinguishes between standard and high priority sectors and there are additional awards for employment creation and for projects in strategic sectors. In Wallonia, a basic award is topped up by complementary assistance for location, job creation, and "essential investment".
Denmark	In March 1985 the base component of the Danish regional incentive package, the Company Soft Loan, was withdrawn and the coverage of the Investment Grant extended beyond the Special Development Regions (SDRs). At the same time, explicit spatial rate discrimination was introduced, the maximum award being 35 percent in the SDRs and 25 percent elsewhere in the General Development Regions.
France	In May 1982 the Regional Development Grant (PDR) was replaced by the Regional Policy Grant (PAT). Under the PDR, three award zones were identified and a distinction was made between setting up and extension. Under the PAT, there are only two zones and no explicit rate discrimination between setting up and extension. However, since July 1987, the PAT has become more discretionary (see Table 2), with particular emphasis on influencing the location decisions of major projects.
Germany	As from March 1981, the maximum preferential rates were reduced by 5 percentage points for all project types except setting up. From 1 January 1983 this reduction was reinstated. From 1 January 1988 explicit rate discrimination between setting up and extension was reintroduced, with rate reductions of 5 percentage points for most extensions and only 2 percentage points for most setting ups.
Ireland	In April 1988 it was announced that capital grants for industrial expansions would not exceed 25 percent (compared with legal maxima of 45 percent generally and 60 percent in Designated Areas).
Italy	Rates of award vary by project size, with higher rates for priority locations and/or sectors. As a result of the 1986 Mezzogiomo Law, the project size quotas increased; in addition, the first implementation plan led to "backward", "intermediate" and "relatively advantaged" areas being designated, with rate discrimination between these areas.
Luxembourg	The Industrial Framework Law of May 1986 introduced explicit spatial rate discrimination into the Luxembourg assistance package for the first time.
Netherlands	In the course of 1980 a three-tier Investment Premium (IPR) system was introduced on a temporary basis (35 percent, 25 percent or 15 percent depending on location). This was made permanent at the start of 1982. In early 1986 rate reductions were introduced for extensions. As from April 1988, the IPR award structure was simplified. There is now no explicit spatial rate discrimination; instead, the main distinction is between setting up projects (and extensions within five years of setting up) and other extensions.
United Kingdom	The British Regional Development Grant distinguished in 1980 between Special Development and Development Areas. After November 1984 the scheme was restricted to Development Areas (Special Development Areas were abolished) and was withdrawn in March 1988. However, there remains spatial discrimination within the UK package. Northern Ireland is treated separately and the key British incentives are restricted to the Development Areas.

Source: Yuill, D. and Allen, K. (eds), European Regional Incentives, various editions (1980-1989).

and, albeit to a lesser extent, fiscal concessions) almost to the exclusion of other policy instruments such measures have tended to be complemented in recent years by "softer" policy measures, relating to technology transfer (patents, licencing) and the improvement of the general business environment through the provision of suitable business infrastructure (including access to sources of advice, consultancy and training).

A number of key developments involving the introduction of "softer" regional policy measures post-1980 are highlighted in Table 5. By and large, the main changes in Table 5 fall into two broad groups. First, in a wide variety of countries, moves were made in the mid-1980s to widen incentive eligibility beyond investment in fixed assets. Thus, in Germany, eligible expenditure was extended to cover the cost of "intangible" assets; in Ireland, assistance was made available towards the cost of obtaining licences or paying royalty charges; in Italy, eligible expenditure was widened to include the purchase of licences, the setting up of offices and the creation of distribution networks; and in Luxembourg, eligible expenditure came to include costs relating to the purchase of "know-how" and technology acquisition in relation to patents and licences.

Second, and of even more significance, important steps have recently been taken in a large number of countries to broaden the regional policy perspective beyond fixed investment grants and physical infrastructure. From Table 5 it can be seen that the 1988 Regional Development Act in Denmark has shifted support away from individual investment projects and towards improving the technological infrastructure of the Danish Development Regions. In a similar vein, a series of regional policy measures were introduced in Germany in mid 1985 aimed to increase aid for the technological infrastructure and provide support for innovative activities. In Ireland, the 1988 IDA policy review has resulted, among other things, in greater regional policy emphasis on new product development and acquisition. The first implementation plan under the 1986 Mezzogiorno Law in Italy has significantly increased funding for innovatory and scientific projects, as well as introducting "planned bargaining" for important innovative projects. The Dutch, have actively considered a "new approach" to encourage innovative problem region projects, though the measures proposed in the December 1987 guidelines were not subsequently adopted. Finally, in Britain a new nationwide, Enterprise Initiative was launched in January 1988 to provide subsidised consultancy support (with higher awards in the problem regions) for specific fields of perceived managerial weakness. In addition, the new Regional Enterprise Grant (introduced on 1 April 1988 for Development Area firms with fewer than 25 employees) provides support for product and process development as well as for investment.

Table 5: Post 1980 Development Towards "Softer" Policy Measures

Country	Change
Belgium	The policy reviews in Wallonia and Flanders in the course of 1987 resulted in more emphasis on innovative activities, activities with a high technology content, and projects of "strategic importance".
Denmark	A new Regional Development Act came into force in April 1988. While the incentives on offer did not change, there was a shift away from direct support for individual investment projects and towards improving the technological infrastructure of the Development Regions. Prior to this, the Regional Development Act of March 1985 had resulted in the withdrawal of the Company Soft Loan and the redirection of funds towards technological development schemes.
France	No major legislative developments.
Germany	In the early 1980s the combination rules for the R&D Investment Allowance and GA Area assistance were eased. In addition, a new Special Investment Grant for High-Grade Jobs was introduced. In mid-1985 a series of measures were introduced aimed, among other things, at aiding technological infrastructure and increasing support for innovative activities. Of particular note, eligible expenditure was extended to cover the cost of "intangible" assets, while infrastructure assistance was extended to include technology-oriented and service- oriented facilities. In addition, a range of new business services became eligible for assistance.
Ireland	The 1986 Industrial Development Act gave scope for introducing new mechanisms to promote development, including the offer of assistance towards the cost of obtaining licences or paying royalty charges. The 1988 IDA policy resulted, among other things, in a greater emphasis on new product development and acquisition.
Italy	The 1986 Mezzogiorno Law extended eligibility for the first time beyond manufacturing to include certain producer services. At the same time, eligible expenditure was widened beyond fixed investment to include the purchase of licences, the setting up of offices and the creation of distribution networks. The first implementation plan under the 1986 Law introduced planned bargaining for important innovative projects in the Mezzogiorno. It also allocated significant new funding for innovatory and scientific projects submitted by public undertakings.
Luxembourg	The 1986 Industrial Framework Law extended eligible expenditure to include costs relating to the purchase of "know-how" and technology acquisition in relation to patents and licences.
Netherlands	The December 1987 regional policy guidelines suggested a "new approach" to encourage innovative projects in the problem regions, though this was not subsequently adopted.
United Kingdom	In November 1984 a number of producer services became eligible for RDG assistance. There also developed a range of consultancy support schemes, largely funded by European Commission monies. In January 1988 a new, nationwide, Enterprise Initiative was launched, aimed at providing subsidised consultancy support in specific areas of perceived managerial weakness. Although national in scope, the scheme offered higher rates of award in the problem regions. Finally, the new Regional Enterprise Grant (introduced on 1 April 1988 for Development Area firms with fewer than 25 employees) provides support for product and process development as well as for investment.

Source: Yuill, D. and Allen, K. (eds), European Regional Incentives, various editions (1980-1989).

On top of such developments expanding the "soft" component of policy, there have, as already noted, been major reductions in the course of the 1980s in the importance of investment-related incentives following the abolition of a number of key schemes most obviously, the Investment Allowance in Germany, the IDA Re-equipment and Modernisation Grant in Ireland, the WIR Regional Allowance in The Netherlands, the Regional Development Grant in Britain and the Standard Capital Grant in Northern Ireland. Clearly, the move away from an almost exclusive reliance on investment-related incentives and physical infrastructure measures towards "softer" incentive measures and developments aimed at improving the business environment generally has been both significant and far-reaching.

From the discussion so far it should be evident that the last decade, and particularly the past three or four years, has been a period of very major regional policy upheaval. While we want to leave a consideration of the main determinants of change until the next section, a key explanatory factor has been the massive reduction there has been, in a number of countries, in regional incentive budgets and spending. In a situation of increasingly limited funding, it is scarcely surprising that regional policymakers have become increasingly interested in coordinating their policy efforts with the activities of other policymakers charged with development objectives. Such inclinations have been strengthened by the move towards non-incentive policy measures and the switch in regional policy focus from the attraction of inward investment to the development of indigenous industry. In such circumstances, the activities of small firm policy, R&D/innovation policy, training policy, urban policy, transport policy and education policy all become of growing significance to the regional policymaker. In a number of countries, this has led to greater attempts to try to coordinate the various development activities in any given area. While it might be a truism that regional policy is about regional development, it is no longer true (if it ever was) to say that regional development is solely about regional policy.

Among the various developments aimed at trying to improve policy coordination, those in Germany, Italy and Sweden are perhaps of most interest. In Germany, the annual regional policy planning document (the Framework Plan) has always stressed the need to coordinate regional policy with other policies of spatial significance (energy, agriculture, urban renewal, vocational training, research and development, environment, the location of government departments and regional planning). In 1985 the Framework Plan made a plea that any new measures in related policy fields "should take into account the situation in the structurally-weak regions at the outset". However, as yet, relatively little progress has been made in this direction, except in respect of a special initiative to cooperate with the Federal Post Office in a joint assessment of telematic provision in the problem regions.

In Italy, an important feature of the 1986 Mezzogiorno Law was the establishment of a system of three-year rolling programmes, revised each year through annual implementation plans. The general aim of these plans is to facilitate policy coordination in the Mezzogiorno by bringing together and integrating the various proposals and programmes of the Southern regions and of central government agencies. Thus, the first plan included budgetary headings not only for incentive policy but also for general infrastructure, the provision of industrial estates, water systems and training. In addition, and as mentioned earlier, a significant new budgetary allocation was provided for innovatory and scientific projects submitted by public undertakings.

Finally, in Sweden, the 1985 Regional Policy Act attempted to increase the degree of co-operation between regional policy and other policy areas with a view to achieving a more concerted promotion of technological change in the problem regions. However, as in other countries, practical difficulties arose in the course of the implementation of the coordination exercise. Nevertheless, the effort has generally been viewed as worthwhile, not least in increasing contacts, information flows and understanding between the various policy areas.

Summing up this section, we have highlighted four main regional incentive policy developments in the course of the past decade. First, there has been a substantial reduction in the importance of automatic incentive instruments and a growth in the significance of discretionary schemes. Second, there has been a considerable increase in the level of selectivity in regional incentive policy, not only through more discretionary incentives but also as a consequence of a notable growth in explicit rate discrimination by area, industry, size of firm and project type. Third, there has been a significant move away from a heavy reliance on investment-related incentives and towards softer policy measures covering licencing, advice, consultancy and training. Finally, in a number of countries, increased efforts have been made to improve the coordination of regional development policy with other development policies, though the obstacles to progress in this area remain severe.

3 An Explanation of Trends

In considering the factors underlying these various trends, it is obvious that a number are primarily country-specific. For example, the abolition of the regional Investment Allowance in Germany at the end of 1988 stemmed from a commitment made during the January 1987 election campaign to reduce government financial support for industry in favour of a reform of the tax system: it was not a consequence of regional policy deliberations. On the other hand, many explanations of change are of far more general significance; and it is on these that this section dwells.

In the previous section, four main regional policy trends were highlighted. In this explanatory section we start with the first of these trends, a strong move away from automatically-administered regional assistance towards more discretionary regional incentive schemes. As already noted, this trend can be attributed primarily to budgetary pressures. In respect of budgets, and taking the Community countries as a whole, the underlying trend has been clearcut. In the latter half of the 1970s there was a major rise in regional incentive expenditure, leading to a peak in most countries towards the end of the decade. Thereafter, decline was marked especially in 1980, but again in subsequent years. While there was something of an upturn in 1986 in most countries, the level of expenditure in that year generally remained below the level of the mid-1970s and was significantly below the peaks of the late 1970s. In the face of the (often severe) budgetary constraints which have been a feature of the regional policy environment in the 1980s, high-spending incentives obviously came under particular scrutiny and, as we have seen, a number of the major automatic ones have been abolished in the course of the 1980s.

However, the move from automatic to discretionary assistance is not solely attributable to budgetary factors. Another very important influence has been the European Commission. In carrying out its incentive coordination function, the Commission (in the form of DG V, the Competition Directorate) has always been keen to sideline schemes which do not fit within its coordination methodology, including certain fiscal measures and a number of grant schemes which were item related (i.e. paid in respect of individual assets) rather than project-related (i.e. determined with respect to overall project spending). Item-related schemes viewed as supporting on-going replacement investment, were a particular anathema to the Competition Directorate. In addition, it has been a general strand of Commission thinking to aim to reduce overall regional incentive spending, causing further pressures to be brought to bear on large-scale automatic assistance. As a result of these various factors, the Commission played a not-insignificant role in respect of the abolition of a number of important automatic regional incentives, in particular Export Sales Relief in Ireland, the WIR Regional Allowance in the Netherlands and the Regional Development Grant in Britain.

A final important explanation of the strong trend away from automatic regional incentives is that the focus of regional policy in the 1980s is very different from what it was in the 1970s. No longer is the attraction of inward investment the prime policy goal; instead, the main policy focus is now on the development of indigenous industry, innovation and entrepreneurship. In such circumstances, the feeling in many countries has been that discretionary assistance is more cost-effective, allowing the incentives on offer to be tailored specifically to the new policy targets and providing more leverage for those projects considered to be particularly beneficial.

The second main trend highlighted earlier was the move towards more explicit rate discrimination in regional incentive award. Combined with the increased discretion already mentioned, there has clearly been a very significant growth of regional incentive selectivity in the course of the 1980s. Once again, budgetary pressures underpin much of this development. In a situation where funds are scarce, there is a strong tendency to focus the available assistance on those projects where it is likely to have most effect. The European Commission dimension has often also been important in leading to increased rate discrimination. The recent introduction of explicit rate discrimination in Germany in favour of setting up projects over extensions, for instance, was largely at the instigation of the Commission. However, increased explicit rate discrimination also reflects the shift in policy focus mentioned earlier away from inward investment and towards indigenous industry and especially towards that part of indigenous industry that is likely to be innovative.

The third main trend identified in the previous section involved the move towards more support for "softer" investment. This development very much reflects the new emphases of regional policy in the 1980s and, in particular, the twin needs to provide an environment in which innovation and technology transfer can prosper and to create conditions under which indigenous entrepreneurship can flourish. But once more the European Commission has played a noteable role, with many of the original initiatives involving "softer" developments being financed by Commission funds.

Finally, we have already made it clear that the recent attempts to increase the level of policy coordination in the regional development field owe much to the budgetary constraints under which regional policy has had to operate in recent years. Financial restrictions on regional policy per se certainly create a need for coordination. In addition, as the regional policy focus has shifted ever more towards the stimulation of indigenous industry, the development objectives of regional policy have become increasingly aligned with those in other policy areas, especially small firm policy, R&D/innovation policy, and policy for education and training. The greater the overlaps of policy the more the potential for coordination. Even so, notwithstanding this increased potential and the growing need for coordination on the part of the regional policymaker, it has to be admitted that, thus far, coordination in practice has been decidedly limited. Other policymakers remain to be convinced of the need for coordination from their (overwhelmingly national rather than regional) perspective.

Summing up, the conclusion must be that the major regional policy developments of the 1980s have been budget driven. Budgetary constraints have been the prime explanatory factor in the move from automatic to discretionary assistance, in the growth in selectivity in award (including more explicit rate discrimination) and in the intensified efforts to introduce meaningful policy coordination.

However, while budgetary factors have clearly been in the policy driving seat, the European Commission has equally obviously played an important role as route-planner. Almost all of the main regional policy developments of the 1980s have fitted in well with the Commission's longer-term objectives, reduced regional incentive spending, less reliance on automatic assistance, more project-based selectivity, increased targetting and more emphasis on innovation and the development of indigenous industry. Moreover, looking to the 1990s and the reformed Structural Funds, it must be the case that the influence of the Commission will grow even more. Indeed, come the 1990s, it will almost certainly be the European Commission which is in the driving seat when it comes to determining the future direction of regional policy with member states relegated very much to a supporting role; always assuming, of course, that the Commission can accommodate itself to the sensitivities of the member states.

4 Conclusions

In this section, we try to identify how the trends highlighted in Section 2 impact on innovative/high tech projects and firms; and what further can be done in order to secure more leverage on these kinds of projects and firms.

In many ways, the trends identified in this paper represent a shift in respect of regional policy instruments such that they are now increasingly favouring, and have greater leverage on, innovative/high tech projects and firms. As we have seen, increased selectivity and rate discrimination have tended, overtly, to favour innovative/high tech firms and projects. The moves towards softer policy measures including advice, consultancy and training also coincide with the needs of many innovative firms where skill levels are above average and thus where training and retraining are key; and where uncertainty and risk can be eased by the provision of information and advice. The fact that, in a number of countries, incentives are being made available to defray some of the costs of the provision of information and advice must, by the same token, help innovative/high tech projects and firms. Similarly, the trend towards deeming intangible assets (e.g. licensing) as eligible for capital grant support must also be a move in the right direction since many innovative projects and firms involve considerable investment in intangible assets. Finally, the growing interest in the need to coordinate regional incentive policy with other national incentive policies (particularly R&D policies), would, if taken further, be of help to innovative projects and firms. In respect of all these elements, then, although one would like to see them taken further, at least the trends are in the right direction.

All of this having been said, there is at least one element in the listing of trends in Section 2 where there must be concern about whether recent developments are favourable for innovative projects, in particular, the move away from automaticity and towards discretion in incentive administration. In addition, and looking to the future, there are other features of current regional incentives which, it could be argued, will need to be changed in order to secure greater leverage on innovative/high technology firms and projects.

We start with the issue of automaticity versus discretion and will conclude on how we think regional instruments might be redesigned in order to give them greater leverage on innovative/high tech firms and projects

In that many innovative projects originate in small firms, the move away from automaticity in incentive administration has been deleterious for such firms which often find it difficult to deal with the complexity of discretionary schemes or accommodate to the uncertainty of award. We, at any rate, have doubts about the cost effectiveness of the very strong shifts there have been towards discretion with its higher administrative resource costs (for awarding bodies and applicants), with the evidence that discretionary incentives are often not incorporated into investment appraisal systems, and with the knowledge that discretionary schemes tend (because of precedents being set) to become largely automatic in terms of rate of award. The uncertainties surrounding innovative projects make discretionary schemes particularly difficult to administer. On the other hand, one can understand the argument that automatic schemes can give rise to large absolute windfall gains. However, if this is the fear, then it can be dealt with (or at least the size of the windfall gains can be much reduced) by using The Netherlands approach, whereby the rate of award of the IPR grant is automatic up to a set absolute amount of eligible investment, and discretionary thereafter. It does seem to us that this approach marries the best of all worlds.

Moving away from this discussion of past trends in respect of financial instruments for regional development, we want to end by considering possible future changes to instruments which could give more leverage on innovative/high tech projects and firms. We want to limit ourselves to four points.

The first arises from the fact that there has been a shift away from soft loans and guarantees as part of the armoury of regional development instruments, and towards grants. It could be argued that innovative projects and firms are, by their nature, of a type which find it difficult to raise capital by normal banking arrangements. Although a grant helps in such circumstances, it still leaves a large funding gap. The problem of raising normal banking capital is especially acute in cases where there are few capital assets to provide collateral against the loan. In these instances, soft loans and loan guarantees are helpful, particularly if the soft loan has interest and principal repayment "holidays" to take account of the longish lead times which many innovative projects have. In brief, the demise of soft loans/guarantees as part of the regional development strategy could be argued to be unhelpful for some innovative projects/firms.

Second, the current regional incentives are still very much capital-oriented. Indeed, the trend over the past decade or so has been away from specifically employment-related incentives and towards capital-related incentives. Yet, many innovative projects/firms are heavily labour cost oriented; and thus the, still strong, capital orientation of current regional incentives means that they exercise little leverage. In such circumstances, these projects/firms would be better helped by incentives which are paid in respect of labour costs, perhaps along the lines of the German Special Investment Grant for High Grade Jobs.

Third, we would question whether the current regional development instruments adequately address a key feature of innovative/high technology firms and projects, namely, the high risk and uncertainty (combined with conditions of high changes) which are associated with them. (High risk and change are not, of course, synonymous. It is possible for a firm to be operating in conditions of high risk and low change; and vice versa.) It must surely be the case, particularly for innovative/high tech projects and firms, that firms are facing greater degrees of change than ever before, whether this be in terms of markets, projects, equipment or technologies. Equipment, particularly, is superceded much more rapidly than in the past and skills very speedily become out of date. But the problems of change are also present in respect of products, where technological progress can, so easily, collapse an "established" market.

These conditions of risk and change argue strongly for consultancy and training to be a key element of regional development strategy and instruments. Consultancy and training enable firms to reduce uncertainty/risk with better information/advice and/or help and to accommodate to changing skill levels. We have already commented that a welcome recent trend in respect of regional policy instruments in Europe has been very much towards strengthening these elements. But we would argue that it is necessary, also, to look again at capital incentives with a view to giving them more leverage in conditions of high risk and change. Repayable grants would be one instrument which could better meet these conditions than a straightforward grant. (A repayable grant is one which is set at a much higher rate of award than current, non-repayable, grants but which is repaid [wholly, partly, or with a premium] in the event of a project being successful.) Such grants are often used by our colleagues in the field of R&D where it has been clearly recognized that risk and change are a dominant feature. Yet, they are barely found in the regional development armoury. Additionally, risk and change often mean short lives for assets. This would, we believe, argue for introducing incentives which have higher rates of award for short life assets than for long-life assets; or, perhaps, there should be a greater reliance on accelerated depreciation allowances. Yet, accelerated depreciation allowances are another regional policy instrument which, over the past decade or so, have become a far less common component of most regional incentive packages.

These, then, are our comments on how regional instruments might be better oriented towards getting more leverage on innovative/high tech projects and firms. There is, however, a final general point with which it is perhaps worthwhile concluding. This is that innovative/high tech firms are by no means uniform. Some have assets such that loan guarantees are irrelevant. Some are capital intensive so that capital grants have leverage. Some are labour intensive and, thus, employment grants have leverage. Some have in-house expertise and in-house advice so they have little need for "state-promoted" resources. Others lack such in-house services. Some innovative high tech firms have capital backing such that repayable grants would not be particularly enticing; others do not. And so on. We do believe, given this diversity, that regional development instruments need to be wide-ranging and varied. It is a point which is relevant irrespective of the target group on which the instruments are focused. It is, however, especially important in the context of the very varied features and characteristics of high tech/innovative firms and projects.

4.2 Financial Transfers versus Real Transfers: Competing Strategies for Regional Economic Development

Manfred Kern

Regional economics started to move in the 1980s, that is regional economic policy as well as regional economic theory. This development has now stabilized far enough to make it possible to speak of a change of paradigm in regional policy.

1 Financial Transfers: Definitional Preliminary Considerations

Up until now the promotion of the regional economy has been pursued according to one single model only, a model that was theoretically and politically fixed and closed for decades: mobility-oriented regional policy. This traditional development strategy basically tried to draw investment capital to regional problem areas or induce settled businesses to make additional investments (Ewers et al. 1980: 2ff.).

Thus, only financial transfers were seen as being capable of helping a region effectively, according to economic policy. The definitions for financial transfer strongly vary in the relevant literature on fiscal economics. According to Dickertmann it is most of all credit aids like loans, debt service aids and suretyships that belong in this category as instruments of financial interventionism (Dickertmann 1980: 32 ff.). But it is sensible for a stricter demarcation towards newer regional political strategies to use a more basic definition, which also includes support by public authorities in the form of subsidies including taxation privileges for private businesses (Semlinger 1987: 277ff.). However, also helpful for the following basic reflections is the bifurcation suggested by Thurow. For Thurow financial transfers are in a broader definition "transfers in cash" and are therefore to be distinguished from "transfers in kind" or "restricted transfers", these will be discussed under the heading of "real transfers" later on (Thurow 1974: 190).

The regional reference of these financial transfers is either made by the federal government (e.g. Gemeinschaftsaufgabe "Verbesserung der regionalen Wirtschaftsstruktur"

(GRW)); (Knigge and Semlinger 1984: 489) or created on the state or county level with appropriate spatially demarcated promotion programmes (e.g. Senate of Berlin 1986).

The goal of financial transfer was (and is) undisputed. The businesses of a region are to be provided with a more favourable liquidity position. This is supposed to put them in a position where they can effectively improve their interregional competitive position and deriving from that secure existing jobs and create new job opportunities. In the light of the traditional regional policy theory this temporary injection of new liquidity is real capital-oriented, which means it will only become active if the promoted business invests in installations and equipment (Ewers and Fritsch 1983: 43). This real capital or investment orientation, results from a factual or assumed regional investment deficit, therefore relative investment deficits, and the resulting disadvantages in production costs, competitiveness and employment. Real capital is considered, following Biehl, as a limiting factor and its assumed limiting influence is to be overcome with a regionally specific financial advancement strategy (Biehl 1976: 62).

2 New Basic Conditions for Regional Policy

Why is this model of financial transfer which has been accepted in the theoretical cause-effect scheme almost unanimously for such a long time, and which has been practiced in regional policy for decades, now being discussed in such an extraordinarily controversial way? Why has the search for alternatives or at least for supplementary instruments, for a purely financial regional promotion been intensified particularly in the last couple of years? Two important reasons seem decisive for this.

2.1 Low Spatial Mobility

First, the relative weakness of growth during the 1980s together with a more strongly noticeable structural change has led to the situation that regional economic promotion especially at the end of the 1970s and the beginning of the 1980s takes place under increasing competition between states and municipalities. Most of all expanding and regionally mobile businesses have become popular objects of these competing efforts. But this is a "critical mass", which has to become smaller and smaller with global growth weakness and rather stagnating markets (Ewers and Fritsch 1983: 56). This diminution of spatial mobility effects the dislocation of already existing businesses as well as the regional canvassing for branch establishments (Bade 1984: 93). Therefore, the confining initial condition, that nowadays one can no longer primarily bet on the settling of new businesses, applies for every municipal economic policy (Cuny 1987: 37).

Against this background, financial transfers have to become less effective instruments for obtaining or even for defending leading positions in the regional competition, irrespective of whether they are carried out as new acquisition or merely as holding on to existing locations. Financial inducements alone lead only to a situation where the regional subsidy volumes are being forced up by every side, without resulting in essential gains from locational decisions of businesses, except for mere windfall gains at relocations already planned anyway.

Wherever single states and communities are successful with this mobility-oriented, financially aggressive regional policy anyway even if it is just temporary, the disadvantaged regions regard this increasingly, freely according to Joan Robinson, as a sort of domestic "beggar-my-neighbour-policy", which costs jobs locally and structurally severs the unemployment problem. In any community that is mindful of political stability, such a policy of regional egoism can therefore not be a permanent maxim of public economic promotion.

Even for a region like Berlin, which due to its geopolitically conditioned artificial island-like location has to suffer under especially great locational disadvantages, a basic change of opinions has taken place in the last couple of years. Lately, the legally fixed Berlinförderung (promotion of Berlin) that was developed to compensate these disadvantages, is being, even though it is still firmly grounded, controversially debated. Criticism not on the principles, but on the kind and the extent of the policy of promoting Berlin can be heard especially from competing regions in the Federal Republic of Germany. They seem to be basically disturbed by the fact that the Berlinförderungsgesetz is based on instruments of financial transfers which achieve this compensating effect. The intensity of this criticism seems to be in reciprocal relation to the amount of promotion given to the region doing the criticizing.

2.2 The Limiting Factor Human Capital

Scepticism with regard to financial transfers also draws on a second, not less important consideration. Many recent regional policy studies seem to indicate that not for a lack of real capital, but lack of human capital the businesses in a region are relatively far behind (Adlung et al. 1980: 14; Ewers and Fritsch 1983: 58; Ewers et al. 1984: 82ff.). Or to put it differently, regional deficits in growth and employment do not especially result from company investment gaps, but from deficits in the qualifications of labour. Qualifications as a limiting factor can refer to the following problem fields: The regional standard of qualifications can be too low altogether, as for example below average quotas of skilled or white collar workers. The number of offers of training opportunities may be too low, since the adequate regional infrastructure has not been developed far enough. In addition, the basis of current experiences with regional qualification offensives it can not be ruled out that the motivation of employees, their willingness for qualification, may be too low. This kind of qualifications deficit can be especially felt in small and medium-sized businesses, the size group that is also most strongly represented in structurally weak regions. But, on the other hand, it is this group of businesses that is considered to be an especially interesting object for regional measures of promotion regarding the desired increase in the number of jobs (Ewers et al. 1984: 35ff.).

3 Real Transfers: Definitional and Preliminary Considerations

A region's poor endowment with human capital is a functional deficit which can not be eliminated through financial transfers alone. Next to the traditional financial transfers a cause-related strategy also has to use the real transfers more effectively, or, in other words, functional deficits have to be replaced, or at least noticeably supplemented, by strategies of real transfers rather than by monetary transfer flows.

These real transfers basically include the immediate provision of services and payments in kind. Material or immaterial transfers of resources that are brought forth in the form of goods or services are known as real transfers."Real transfers mean material and immaterial transfers of resources, of goods as well as services" (Semlinger 1988: 7). Real transfers are supposed to contribute directly to the elimination of weak points in companies, improve the use of business resources and speed up the organizational dynamics of growth and development (Semlinger 1988: 7ff.). Real transfers in the sense of in-kind payments e.g. in the form of a direct provision of machines or buildings are thereby an alternative that is rather theoretical and less practicable for regional policy. Therefore, it will not be dealt with any further.

According to the experiences so far, the possible comparative advantages of the real transfers over the financial transfers seem to lie more in the field of services. This main point again can be divided into pure transfers of information and advice (transfer of knowledge) and into concrete transfers of personnel, which of course always and necessarily include a transfer of knowledge. Often in the literature the transfer of technology is also discussed as an independent variant next to the transfer of information and the transfer of personnel (Allesch 1987: 99ff.). Most of all the technology transfer agencies in their function as mediator between medium-sized businesses seeking advice, on the one hand and experts in science and the economy, on the other, are addressed (Allesch 1987: 100). But since on this level of transfer information is given to businesses, either directly, as transfer of knowledge or through persons, as transfer of personnel, a distinct presentation of technology transfer can be done away with due to the considerations made here.

4 Models for Real Transfers

Measures of real transfers still have a distinct experimental character in present regional policy (Semlinger 1988:70). The discussion about the merits and weaknesses can therefore not yet ensue on the basis of a closed conception of regional policy, but only as a recourse to experiences from single examples and models , that are momentarily in the state of being tested (Semlinger 1987: 279ff.).

Meanwhile a comparably broader knowledge of the transfer of information and advice exists, since it has made its way on the regional level everywhere in the EEC for a long time now. Institutionally this transfer level, most of all, includes all the regionally very different, but altogether extraordinarily broad spectrum of universities and colleges, federations, chambers, and corporations for the promotion of the economy and special technology transfer agencies (cf. Chris Hull, in this volume).

5 Transfer of Knowledge Through Persons in the Berlin Economy

Concrete examples for experiments in regional policy with the transfer of personnel can be found partially original, partially pars pro toto in the Berlin economy. Three approaches are especially interesting in this context, since they can, at the same time, be shown as a sort of instrumental step model of personnel transfer. A criterion for the classification of the different steps is thereby the degree of liability with which the knowledge can be brought to the businesses and transposed on location.

5.1 Consultants for Training and Advanced Training

Consultants for training and advanced training should contribute to raising the qualifications level of the trainees and other employees in the businesses. The Chamber of Industry and Commerce in Berlin employs nine education consultants, who altogether make about 5,000 company visits. Six advanced training consultants, who as part of a pilot project are being financed with public funds, have visited more than 130 businesses and have initiated 33 internal measures for advanced training in 1988. In the sense of the step model mentioned, both main points of consultations education consultation and advanced training consultation are only offers to the businesses. Even though the advisors visit the companies, they do not put their suggestions into effect themselves, but rather leave the decision on this subject to the advised businesses. Education and advanced training therefore lie on the borderline between the pure transfer of information and the information-based transfer of personnel.

5.2 The Innovation Assistants Programme

The innovation assistants programme provides a pertinent example for the second step of the transfer of personnel (Senat von Berlin 1986: 68ff.). It exists in Berlin since 1982 and its main characteristics have meanwhile been taken over as the so called Berlin model by several other states in the Federal Republic of Germany, for example Bremen, Lower Saxony and North-Rhine Westphalia (Allesch 1987:105). Innovation assistants are especially qualified, freshly certified university graduates, employed by medium-sized businesses. They are supposed to take away the possibly existing threshold fear of academically trained personnel and give the respective company access to innovative problem solving that could not have been tackled before, due to a lack of sufficient specialized qualifications of the available employees. Businesses with a turnover that did not exceed 100 million DM in the previous year that hire an innovation assistant receive a 40% subsidy by the Senate to the gross income, liable in wage tax for one year (24,000 DM maximum). Altogether 655 innovation assistants where brought to medium-sized businesses in Berlin in the past six years (as at October 31, 1988). There was an especially high demand for graduates of courses in electrical engeneering, precision mechanics, engineering and computer science. The field of technical environmental protection is developing a growing demand. More than 57% of these innovation assistants are working in companies with up to 20 employees; 31% are working in companies with up to 100 employees. Most of the assistants stay in the businesses longer than the sponsored first year. We can derive from this, that the assistant programme has contributed to an increase of employment in the small and medium-sized businnesses of the Berlin economy.

5.3 The Medium-Sized Business Assistant and the Production Assistant Programme

From a methodological viewpoint the personnel-transfer models of the medium-sized business assistant and the production assistant are the ones furthest developed (Senat of Berlin 1986:70,79). In both cases a phase of advanced training precedes the activity in the company (12 months for medium-sized-business and 18 months for production assistants).

5.3.1 Medium-Sized Business Assistants

Medium-sized-business assistants are graduates of a business or technical college. They are supposed to relieve the executive personnel of medium-sized businesses of their daily work and give them more room for strategic and conceptional thinking. The main points are the fields of marketing and sales, cost control and finance and the training schemes. Promotion is basically oriented at the model of the innovation assistant. So far 20 to 25 medium-sized business assistants have been trained in each of two groups. Funds for a third group have already been granted. More than 90% of those trained with the first group found positions in small and medium-sized companies.

5.3.2 Production assistants

Production assistants are recruited from the social sciences i.e. education, political science and sociology. They are supposed to contribute their specific expert knowledge in industrial development projects and production processes within an interdisciplinary approach. More concretely, they are supposed to help faciliate the introduction and adaptation of new production techniques in a socially acceptable way.

The programme "production assistant" began in December of 1986 with 15 participants; a second run started in autumn of 1988. Of all the measures of regional transfer of personnel that have been sketched here, the production assistant certainly has the strongest experimental character. For a positive reception by the businesses it could be decisive that this test run in Berlin is being supported jointly by the university (Free University of Berlin) as well as by a highly reputed high-tech institute (Fraunhofer Gesellschaft and Institut für Produktionsanlagen und Konstruktionstechnik Berlin).

6 Relative Weight of Financial and Real Transfers

This listing is neither complete nor fixed in stone as its examples have, as already stressed, a strongly experimental character. This indicates that a regional political deus ex machina for an accelerated mastering of structural change is not yet available with regard to real transfer. Furthermore, with the favoured introduction of the transfer of personnel, it is not intended to introduce this variant of real transfer as an alternative that could fully replace the classical instruments of financial transfers. For one thing, this would not work, because naturally the real transfers also have to be financed, partially even in the problematic form of direct labour cost subsidies (these being fixed in time and costs!).

Furthermore, the politically conditioned isolation of Berlin means that under aggravated structural conditions a regional political compensation of disadvantages can not dispense of financial compensation measures. Therefore, the policy of promoting Berlin (Berlinförderung) is a special form of financial transfer that is indispensable for the economy of the city and can be conveniently supplemented through personnel-supported transfer models, but in no way substituted. Similar considerations also basically work for the financial compensation of politically conditioned locational disadvantages, as is the case for the regions bordering with the German Democratic Republic.

7 Change of the Regional Political Paradigm

Therefore, it would certainly be exaggerated to speak of a change of regional political paradigms in the immediate comparison of financial transfers and real transfers. If such a change actually is going on at the present time, then it is going on at a higher and here this means a more conceptional level of regional economic promotion.

What was in the beginning circumscribed as a paradigm of mobility-oriented regional policy is now, indeed in a sort of relative state of descent. This development has occurred mostly because mobility-oriented strategies mainly rely on instruments of financial transfers and therefore try to bind external resources, especially investment capital, to their particular regions. The distinct decrease of effectiveness in all financial measures that are supposed to trigger or support spatial mobility corresponds with the retrogression of this external potential.

On the other hand, the obvious alternative, to concentrate on the region's own potential and to better promote that, is winning adherence, theoretically as well as politically. This prefered promotion of the internal or indigenous potential through innovation-oriented regional policy has meanwhile acquired such a clear shape that one can by all means speak of a new regional political model (Ewers and Fritsch 1983: 56). Most of all, numerous new forms of real transfers are instrumentally included in this model, in addition to existing financial transfers. Therefore, the potential of real transfer offers should not be underestimated for the present as well as for the future regional policy, even if it can momentarily only be proven experimentally (Semlinger 1988:314).

8 Promotion of Competition Through Real Transfers

The fact that real transfers can be offered, at least basically, over markets and, therefore, privately even where this is only brought on its way via federal impulse-financing, speaks for this estimation (Semlinger 1987: 281). The growth markets data banks or management consulting firms can be used as proof, as can the ambitious agglomeration of consultation services in founder, technology and lately also in service centres, and especially in such centres where the setting-up is financed in whole or at least partially by private enterprise.

Financial transfers, on the other hand, are basically not privatizable, since they are by definition government expenditures (or public income-wavers). One point in favour of a characterization of real transfers as a private good is the fact that it can bring the competition between regions back to its "natural" basis. Such regional economic policy concentrates on real transfers, utilizes the existing (and therefore "natural") resources of the region and tries to transfer them into comparative advantages to the other regions. From this point of view it becomes especially obvious, how important a broad spectrum of research and development activities are for a region. They offer decisive suggestions for the use of real resources which are necessary for a successful transfer of personnel and can be used in the regional competition as action parameters.

The competition-promoting qualifications of real transfer are also expressed in the fact that this concept can be realized in a supply-oriented as well as in a demand-oriented way. Its aim is to strengthen the suppliers (or supporters) of real transfers, especially including the local universities, colleges, research institutes and mediating transfer units. At the same time this transfer concept addresses demand (or users) especially small and medium-sized businesses to improve their qualifications, to increase their growth potential and therefore their potential labour force and thereby consolidate their position in the interregional competition by drawing on the indigenous potential of the region.

Before the background of the European single market, which will come into effect in 1993, it should finally be said that this concept of personnel-supported transfers of know-how would, because of its spatial footholds, also excellently adapt to a European model of regional policy. This applies most of all, when European-wide regional goals are aspired to promote the indigenous potential of a bigger, competitively structured economic area trough innovation-oriented regional policy.

References

Adlung, R. and C. S. Thoroe (1980)
Neue Wege in der Regionalpolitik. Vorschläge zu einer Umgestaltung des regionalen Fördersystems in der Bundesrepublik Deutschland, Kieler Diskussionsbeiträge, Nr. 68, Kiel.

Allesch, J. (1987)
Die Bedeutung des Technologietransfers für industrielle Klein- und Mittelunternehmen. Erfahrungen aus der Praxis. In: Bundesverband der Deutschen Industrie (Hrsg.), Innovationen Zukunftschancen für den industriellen Mittelstand, Köln, S. 97–109.

Bade, F.-J. (1984)
Industrieansiedlung in der Bundesrepublik Deutschland, in: Deutsches Institut für Wirtschaftsforschung, Wochenbericht Nr. 8, S. 92–98.

Biehl, D. (1976)
Bestimmungsfaktoren des regionalen Entwicklungspotentials Lage, Verkehrsinfrastruktur und Umwelt als Anwendungsbeispiele des Potentialfaktorkonzepts. In: Die Weltwirtschaft, Heft 1, S. 60–65.

Cuny, R. H. (1987)
Die Bestandspflege rückt in den Vordergrund. In: Wirtschaftsdienst, S. 34–39.

Dickertmann, D. (1980)
Öffentliche Finanzierungshilfen, Baden-Baden.

Ewers, H.-J., R. Wettmann, J.F. Kleine and H. Krist (1980)
Innovationsorientierte Regionalpolitik. Internationales Institut für Management und Verwaltung, Wissenschaftszentrum Berlin, Berlin.

Ewers, H.-J. and M. Fritsch (1983)
Beschäftigungswirkungen regionaler Wirtschaftspolitik. In: D. Garlichs, F. Maier, K. Semlinger (eds.): Regionalisierte Arbeitsmarkt- und Beschäftigungspolitik. Frankfurt am Main, New York, S. 38–65.

Ewers, H.-J., M. Fritsch and J. Kleine (1984)
Bildungs- und qualifikationsorientierte Strategien der Regionalförderung unter besonderer Berücksichtigung kleiner und mittlerer Unternehmen.
Schriftenreihe 06 "Raumordnung" des Bundesministers für Raumordnung, Bauwesen und Städtebau, Nr. 06.053, Bonn.

Knigge, R. and K. Semlinger (1984)
Beschäftigungsorientierte Reform der regionalen Strukturpolitik.
Von der Sachkapital- zur Humankapitalförderung.
WSI-Mitteilungen, Heft 8, S. 487–495.

Semlinger, K. (1987)
Qualifikation und Qualifizierung als Ansatzpunkte beschäftigungsorientierter Strukturpolitik: Finanzhilfen versus Realtransfers.
In: Fritsch, Michael, Christopher Hull (eds.): Arbeitsplatzdynamik und Regionalentwicklung. Beiträge zur beschäftigungspolitischen Bedeutung von Klein- und Großunternehmen, Berlin, S. 267–289.

Semlinger, K. (1988)
Staatliche Intervention durch Dienstleistungen.
Funktionsweise und Steuerungspotential untersucht am Beispiel der technischen Beratung zur Förderung der betrieblichen Integration Behinderter, Berlin.

Der Senat von Berlin (1986)
17. Bericht über die Lage der Berliner Wirtschaft und die Maßnahmen zu ihrer Weiterentwicklung, Berlin.

Thurow, L. C. (1974)
Cash Versus In-Kind Transfers. The American Economic Review, Papers and Proceedings, Vol. 64, 190-195.

4.3 Information and Consulting as Instruments of Regional Development Policy: The Role of Public Agencies

Christopher John Hull

The trend in regional development policy towards strategies for "mobilising indigenous potential" has led to a particular focus on small and medium-sized enterprises (SMEs) - on encouraging the formation of new firms and the growth of existing ones. Many policy-makers have embraced the argument that one of the preconditions for a flourishing small-firm sector in a region is an adequate minimum supply of essential business-support services.

It is in this context that there has been marked growth in locally provided information and consultancy services in the last few years. The range of these services is broad: business plan preparation, financial management, technology auditing, market research, financial intermediation, project management, technology transfer, marketing strategy, etc. The providers are also varied: public (regional development agencies, local governments, universities and technical colleges, etc.), parapublic (chambers of commerce, local development associations) and private (commercial consultancy firms). Many initiatives are joint ventures involving both public and private actors.

It was originally intended that this paper should attempt an empirical typology of both the various kinds of information and consultancy services currently offered to firms in Europe and of the different sources of their supply. The idea was that a cross-tabulation of the two might yield some insights into the proper role of public agencies in the provision of such services. I have decided against attempting this. The situation in Europe is too fluid, and poorly documented, to make such an analysis feasible without a major prior research effort. Instead, I shall consider some broader principles relating to the public role in the provision of information and consultancy services to firms. In particular, I shall ask whether the public sector should itself supply services (direct provision) or whether it should rather concentrate on an intermediary role aimed at encouraging others to supply them and firms to use them.[1]

1 SMEs Need for Information and Consultancy

Successful economies and businesses are built on right decisions about products and markets. Those decisions are founded on factual information and judicious interpretation. In general, the problem of information is less that it does not exist than that high costs are involved in its generation, acquisition, analysis and interpretation.

Those costs are, by and large, relatively unproblematic for large firms, which are able to devote significant resources to "information-management" functions. Most SMEs lack the means to do likewise. For this reason, perhaps, many studies have indicated that poor management, i.e. inadequate decision-making, is the principle reason for the failure of SMEs. The implication is that better management would help SMEs better identify and exploit opportunities for growth; new firm formation rates should increase and failure rates decline.

In principle, SMEs could compensate for their lack of information- management capacity by making use of externally supplied information and consultancy services. In practice, relatively few resort to this option - certainly in a systematic way.

2 Factors Inhibiting SMEs' Use of Information and Consultancy

Lack of space precludes a long discussion of the reasons why SMEs make little use of information and consultancy services. For present purposes it may suffice to identify four: availability of services, their cost, ignorance as to their availability, and firms' myopia as to their own problems.

2.1 Availability

It is evident that many regions are poorly endowed with business services of all kinds. Generally speaking, the further down the urban hierarchy one goes, the worse the provi-

---

1 Many of the arguments in this paper are set out in greater length in:

Christopher Hull with Benny Hjern, *Helping Small Firms Grow: an Implementation Approach*, Croom Helm, London, New York, Sydney (1987). The interested reader is referred to that source, in particular chapters 1 and 9, and to the literature cited there.

sion. In peripheral areas, there may be no more than a branch office of a savings bank and a general-purpose lawyer.

Clearly, services do not need to be locally based to be locally accessible. Services based in market towns, county capitals and major cities do serve wider hinterlands. However, the costs of use generally increase substantially with distance.

2.2 Cost

Even where services are available, SMEs are often reluctant to use them for reasons of cost, real and perceived. It is one of the hoary old truths of the consultancy professions that nothing is as hard to sell as information. When a firm buys a machine, it knows what it is going to get when it signs the contract of sale. When it signs a consultancy contract, by contrast, it commits itself to a certain cost for an often uncertain return: there is no guarantee that the consultant will provide a useful solution to the firm's problem. While, objectively, such uncertainities can be reduced by, for example, the careful selection of consultants, the precise specification of the consultancy assignment, and, more rarely in practice, a payment-by-results formula, many SMEs remain reluctant to use consultancy and information services because of the uncertain cost-benefit relationship.

2.3 Ignorance

It has to be assumed that many small firms are simply unaware of the consultancy services which are available, or have only a vague idea of what particular types of consultant can (and cannot) provide. This is one of the key problems which any public policy of business-service provision has to address.

2.4 Myopia

Very many small firms do not act strategically. Their owners and managers are completely absorbed with the day-to-day running of the business. They fail to see problems looming on the horizon, and so lose the chance of tackling them before they assume crisis proportions. Even when problems are apparent, firms are often unable to identify their precise causes, appreciate their ramifications, or structure them for solution. At one level, this is simply to state again that small firms have an objective need of the services of the external consultant, whose job it is to identify problems, analyse them, and propose solutions. In practice, however, the effect of such myopia is to prevent or discourage many firms from seeking external assistance.

Clearly, the firm which fails to see problems looming on the horizon has no motivation to seek the services of a consultant: it does not yet know that it has a problem. More perverse can be the reaction of the firm which realises it has a problem but is unsure of what it is, or how to tackle it. The reaction is often to wait in the hope that the problem will assume clearer shape, or will simply go away.

Psychological factors are often at the root of such reactions. It can be difficult for a small-firm owner, whose firm is his life and his life's achievement, to admit that he has lost control. Admitting a problem that he cannot comprehend and tackle can amount to admitting failure. It is perhaps this more than anything which causes small firms with problems to close in on themselves rather than to go out and seek external assistance.

3 Roles for Public Agencies in the Supply of Information and Consultancy Services

The foregoing analysis suggests three possible roles for public agencies in the supply of information and consultancy services:

- Direct provision of services, notably where they are unavailable
- Subsidisation of services, to increase marginal supply
- Activation of demand, to encourage firms to make greater use of services

I will discuss each of these in turn, and then suggest a more comprehensive "intermediation" strategy which incorporates elements of all three.

3.1 Direct Provision

This option may be appropriate wherever services which can be considered critical are currently not available. This will be true of many kinds of services in rural and peripheral regions, in particular, and possibly of certain more specialised services in central locations. Economic efficiency arguments can make the decision whether to provide services directly both objectively difficult and politically divisive. I do not propose to consider those, often convoluted, arguments here, as important as they are.

I will, however, raise several other points which merit attention whenever direct public provision is being considered, and which are perhaps not immediately obvious. One is that it will generally take many years to build up effective local information and consultancy services and achieve a significant impact on firms. It is therefore important to be able to count on a stable medium- to long-term public commitment if the initial investment is not to be wasted. Where politics are partisan and majorities likely to change, this may not be realistic.

A second and related point is that public sector financial planning may not be structured so as to allow clear longer-term commitments. Where budgeting is annual or short-run, and where budgets are tight, it may not be wise to commence a necessarily longer-term investment in developing services.

Third, if public service remuneration regulations apply, it may be difficult to attract and retain high-quality personnel. Good consultants, for example, will usually expect a total salary package which equates with what they could earn in private-sector employment, which will often be more than the public sector is prepared to pay. The danger is high turnover in key personnel, with consequent damage to the development and quality of services.

Where these kinds of constraints seem real, the better strategy may be to resort to the subsidisation option (see below) with the aim of increasing the supply of commercially provided services. I would add that it is also important to apply mechanisms which ensure that publically provided services, where instituted, are responsive to the real needs of firms. The simplest of these is the price mechanism. In other words, firms should be made to pay something, although not necessarily the full cost, for services.

3.2 Subsidisation

Subsidies can be used to stimulate the supply of commercially provided services, both to encourage the expansion of existing services as well as the creation of new ones. In general, the aim will be to apply subsidies at the margin, so as to reduce costs by the minimum amount necessary in order to make the expanded or new service commercially viable at some targeted level of activity.

Subsidies can in practice be targeted at users or providers. Both approaches may be necessary. Particularly where a new service has high development or launch costs, subsidies will need to go to the provider. Where, by contrast, the aim is to expand the marginal supply of an existing service, by overcoming firms' price resistance, it may be preferable to subsidise via the users, by reimbursing part of their costs of using the service. The reason for preferring user subsidies is, again, to ensure that services are attuned to firms' needs.

3.3 Activation of Demand

In principle, it can be expected that the availability of subsidies will give service-providers a powerful marketing tool. Because they can offer their services at lower prices, they will have an incentive to seek out new client firms. Whether and to what extent this will happen in practice, however, is difficult to estimate.

The costs of marketing such services to a large and diverse population of SMEs can be high. It may need a substantial investment in, especially, personal visits to many firms before any substantial new business is generated. One response could be to subsidise suppliers' marketing costs, in addition to subsidising the price charged to users.

Subsidies can be expected to work where the principal obstacle to service up-take by firms is price resistance. They will be neither effective nor efficient where the real obstacle is a "cultural" resistance to allowing "strangers" into the firm, or myopia as to the existence of problems in the firm. Such obstacles can only be tackled by someone with privileged access to the firm's owner, someone who is trusted by him as a both competent and neutral adviser. Commercial service providers will rarely be able to achieve this relationship with a firm, for economic reasons and by virtue of their own partiality.

4 Local Agencies and Intermediation

Circumstances obviously vary from place to place but I would argue, generally, that the fundamental obstacle to the greater use of external information and consultancy by SMEs tends to be the kind of cultural barriers to which I have just referred.

In consequence, and given the improbability of commmercial service suppliers being able to break those barriers other than extremely slowly and selectively, I shall argue that this should be a key role for local agencies.

The challenge is to gain initial access to firms, and to gain their trust. This can only be achieved by a gradual process of confidence building aimed at reaching the least introverted firms first, and at extending the outreach to the more reluctant ones later - by demonstration, recommendation and gentle persuasion. It has to be admitted that not every public agency and public servant will have the "image" and competence needed to achieve this. Serious thought needs to be given to their credibility vis-à-vis the target firms. It may be that the public agency should take the initiative but remain in the background, using firms that are opinion leaders and perhaps other institutions (e.g. employers' associations) to relay the message. Similarly, sensitive marketing mechanisms will need to be devised (e.g. publicity featuring positive-case firms, introductory seminars with target firms on "neutral" subjects) in order to excite interest and establish first contacts.

Thereafter, the role of the well-established intermediary will in effect be to match demand for and supply of information and consultancy services in his territory, in a more or less hands-on way. Where services are well developed, his role may be limited to activating marginal demand among firms and to putting them individually into contact with the specialist information and consultancy suppliers able to meet their specific needs. In some cases he will need to go further, helping firms to negotiate and manage consultancy assignments. This will apply in particular to firms gaining their first exposure to external information and consultancy.

Where services are under-supplied, or where there is residual price resistance among firms, subsidisation strategies may be brought into play. I suggested earlier that subsidies should, where possible, be paid to users, and I would argue further that intermediaries should make the awards to them. The advantages are that the availability of subsidies via the intermediary is likely to increase his chances of motivating firms in the first place as well as to ensure that subsidies are targeted at firms with problems having a realistic probability of being resolved through external assistance. This strategy does, admittedly, suppose intermediaries of high competence.

There are potential disadvantages in this kind of hands-on intermediation strategy. One is that the available information and consultancy suppliers will almost inevitably have to be vetted in some way for quality, in order to avoid recommending and subsidising the ineffective. Another is that subsidies awarded to users entail higher administrative costs than subsidisies given to a smaller number of suppliers. But where the awards are well targeted by the intermediary, the gains in efficiency and effectiveness should outweigh the costs.

An additional advantage of this kind of intermediation strategy is that the well-established intermediary will necessarily generate a sound knowledge of firms' potential demand for currently under-supplied services, which can be used to encourage potential suppliers.

In localities where particular services are unavailable, the intermediary may consider providing them directly himself. Alternative strategies, involving less commitment by the intermediary, include joint ventures with existing suppliers, subsidies to new suppliers to help them develop and launch services, or contracting with suppliers outside of the region to make services available, e.g. on a regular "surgery" basis. There is a broad range of possibilities.

This relatively hands-on role for local public agencies as inter-mediaries is not uncontroversial, but especially in localities with an economic structure which predominantly comprises traditional SMEs, and where the current supply of services is inadequate, I can see little realistic alternative if one wishes to encourage greater use of external information and consultancy by firms. Simply to supply services directly, or to promote them through subsidy, is unlikely to succeed without prior attention to the willingeness of firms and ability to make use of them. The pre-requisite, however, is a longer-term political commitment to the strategy.

Reference

Hull, C. J. with B. Hjern (1987)
Helping Small Firms Grow: An Implementation Approach.
Croom Helm, London - New York - Sydney

4.4 Improvement of Regional Qualification Structures as a Task of Regional Economic Policy

Benny Hjern

1 Introduction

There are 284 municipalities in Sweden. In 1987, a government commission identified at least 515 "resource centers" throughout the country; mainly nonprofit organizations established to provide some kind of technical competence to qualify (aspiring) entrepreneurs, firms and post-gymnasium students. Among the centers identified some 80% had been established since 1980. Of those less than one third were affiliated with universities/colleges or other directly state-funded institutions and another 15% with municipal technical high schools. More than half of the centers claim a rather mixed sponsorship; banks and firms together with municipalities and regional state agencies as well as (mainly state funded) regional venture capital organizations.

To non-Swedes, the flourish of local activity throughout the 1980s to establish local resource centers, may lead to the conclusion that, surely, in centralized Sweden, this surge must be due to a new state policy to improve regional qualifications structures. Such a conclusion would be in line with the proverbial model of Sweden (presented to foreigners in government script). But it misses the politico-administrative changes that have evolved in the central-local set-up of regional policy. These changes make it more difficult than before to analyze the "policy" for improving regional qualifications structures.

It should be stressed, therefore, that presently an account of the improvement of regional qualifications structures in Sweden can hardly rely upon the notion that it is the result of well-integrated and articulated regional policy. Currently, it is rather the result of the efforts of regions and municipalities in search of a national policy, manifested in local projects of varying focus and ambition which are still experimental as far as forms of management and control are concerned.

Twenty years ago, most of the discussion on regional qualifications structures could have concentrated on the training aspect of Swedish selective labor-market policy. The bulk of the changes in regional qualifications ensued through labor-market training (AMU) centers. The effects, however, were not necessarily related to the improvement of qualifications structures in underdeveloped regions, since the demands of national rather than regional labor-markets were stressed. The AMU centers are still part of the qualifications set-up, but now with a much more local, indigenous labormarket focus. However, so are the universities/colleges, municipal high schools, private firms and consultants to a much higher degree than ever before.

Improvement of regional qualifications structures, in the 1980s, has evolved into something like a growth industry in Sweden. Information on the policies and institutions shaping up is patchy and scattered. For example, the rapid establishment of "resource centers" in the 1980s forms part of a trend towards municipal mobilization in the field of local industrial policy. This more active involvement of the municipalities, however, has not yet been adequately documented and analyzed. But the surge does indicate a shift towards local rather than regional mobilization, and that, at least in theory, state regional policy has recently started to focus its instruments in that direction.

In discussing who takes on the task of improving regional qualifications structures in Sweden and with what results, it is neither sufficient to concentrate on the regional policy administration proper, or its instruments, nor is it satisfactory to consider only the new technical "resource centers" - although they do seem to involve several of the actors deemed important in Swedish regional policy. Qualifications strategies engage assorted associations of Swedish actors with truly mixed-economy fund-raising capacities.

This article uses materials on the strategies of municipalities and others involved in establishing resource centers as one empirical focus to shed some light on the new institutions forming in Sweden to take on qualifications tasks. But it also provides materials on the role of universities and municipal high schools in upgrading technical skills in Swedish regions.

The discussion commences with a broad overview of trends in the demand and supply of human capital investments in Sweden and information is provided on regional qualifications. A brief account is given of the place of qualifications strategies in the formation of state regional policy from the 1960s onwards. Subsequently the role of key actors in Swedish regional qualifications structures is described and, in conclusion, analyzed.

2 The Qualifications Setting

The Swedish labor force is considered to be one of the best trained in the world. Still, Swedish investment in human capital, as measured by the share (%) of R&D of GNP, increased rapidly in the late 1970s, and since 1985 exceeds that of the FRG, Japan and the US (Fig.1). One explanation given for the trend is the increased awareness of and need for structural changes pinpointed by the prolonged crisis of traditional Swedish industries in the 1970s. The importance of "knowledge" as a factor of production has been buttressed (cf. Eliasson 1985).

During the 1970s and early 1980s the proportion of the labor force engaged in "knowledge production" had already increased rapidly. In the 1960s, some 10% of the employed were engaged in tasks related to research, teaching, consulting, and information processing. This share doubled in the 1980s and is expected to triple during the 1990s. From the middle of the 1970s to 1985, R&D-intensive firms (more than 15% of value added invested in R&D) in manufacturing increased their share of total value added in the industry from 7% to 25% and this is forecast to increase to 50% by the turn of the century (cf. Andersson 1988:180 f). Today manufacturing, in expenditure per man-year,

Fig. 1: R&D expenditure in percent of GNP in OECD countries

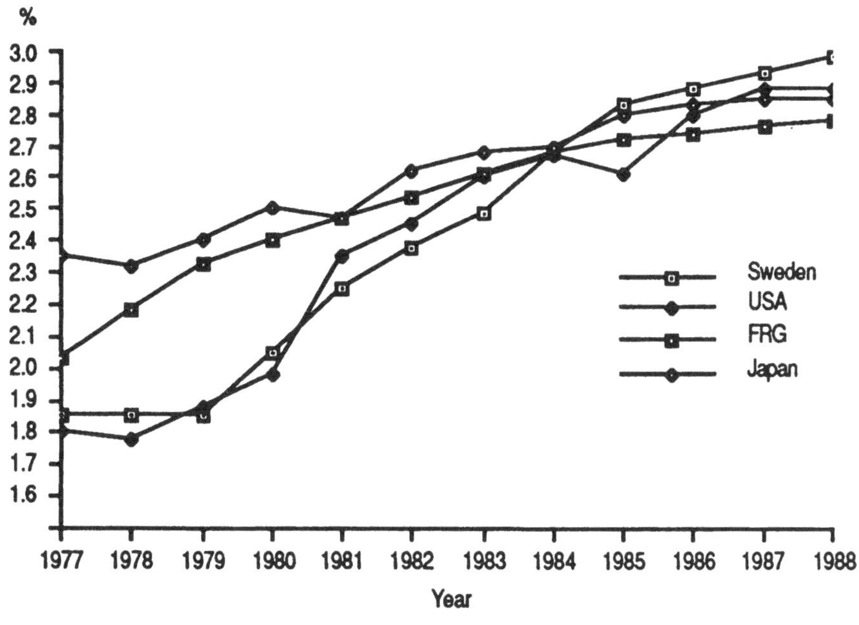

represents 75% of private sector R&D, but large firms and some branches e.g. transportation/communication, electronics and machinery heavily dominate R&D developments in the sector (Ekstedt 1988:40f). In manufacturing, "soft" investment costs have increased faster than other types for quite some time. In 1985 these costs were 50% larger than for other types of investments (Swedish Central Bureau of Statistics, Survey 1986).

Presently, qualifications measures in Sweden are rapidly being upgraded and redirected to adapt to new demands of industry (and public agencies, although detailed information here is scarce in view of what is demanded by public organizations). The bulk of R&D investments pertain to firms (see Table 1). In 1985 some 35% of total R&D man-years were in the public sector, mainly in universities and technical institutes, meaning that the share of the public sector in total R&D investments is lower in Sweden than in comparable countries (Ekstedt 1988:41).

R&D trends just indicate in very broad terms the pace of change at which the qualifications of human capital are demanded by the production processes of the economy. The dearth of detailed data on actual demand and supply of qualifications measures in Swedish firms and public agencies allows very general conclusions only: recently demand has risen and the pace of change has quickened, and it can be surmised that the demands of firms increasingly are supplied by providers other than universities and colleges. But relevant courses of labor-market training have decreased rather than increased their number of trainees throughout the 1980s (National Labour Market Board. Labor Markets Statistics 1988:7). Thus, it is likely that (foreign and domestic) consultants and the new "resource centers" have increased their share of the qualifications markets.

Table 1: R&D man-years in branches of the Swedish economy (percent)

Year	Private sector	Public sector (without universities)	Universities and colleges
1975	67	10	23
1977	67	10	23
1979	67	10	23
1981	62	12	26
1983	63	10	26
1985	66	10	24

Source: Ekstedt 1988: 50

3 Trends in Qualifications Structures

During the late 1970s and early 1980s investment in human capital increased rapidly in Sweden. By the end of this period it is also known that the urban areas, and especially the Stockholm region, had started to attract labor again, after the earlier "green wave" period of labor-market migration. What were the effects on regional qualifications structures of these trends?

In Sweden the term "forest counties" is a regional economics euphemism for the underdeveloped regions. From 1971 to 1983 a slight decrease[1] - from Gini index 1971 = 10.5 to 1983 = 9.1 - is actually documented for the differences between "forest" and other Swedish counties (Åberg 1988: 200ff). Hence, notwithstanding the fact that investment in human capital started to soar during this period, data indicate that inequalities in regional qualifications structures did not increase as a consequence.

If a similiar, crude test is applied only to the manufacturing sector of the Swedish economy, which invested more in human capital than any other sector in the period 1971-1983, data suggest that, on balance, differences in regional qualifications structures were not much affected in that industry either (see Fig. 2).

On the one hand, the proportion of employed with just pre-gymnasium training decreased more in the forest counties than elsewhere in Sweden, but, on the other, the proportion of post-gymnasium employees increased more in the large cities than in the forest counties. Manufacturing firms in the the latter had very few employees with post-gymnasium training in 1971, and 4% in 1983. The large cities increased their proportion from 4% to 13% during the period.

Data seem to indicate that firms in the forest counties, in the period 1971 to 1983, more or less have been able to attract the same proportion of qualified employees as elsewhere in Sweden; possibly with an exception for labor with post-gymnasium training. A well-educated guess would be that the latter difference pertains more to large firms, which have concentrated tasks requiring greater skills, than to headquarters establishments in one of Sweden's three major cities (cf. Ekdahl and Lindström 1985).

A direct comparison between qualifications structures in Sweden's forest and large city counties with regard to types of firms is not possible, due to lack of data. A study comparing the most northern Swedish forest county (Norrbotten) and a southern, more industrialized one (Jönköpings län), provides some additional insight on differences of regional qualifications structures, however (Sundin and Wiberg, 1985). The former county rep-

Figure 2: Formal training of those employed in manufacuring in 1971 and 1983. Percent with pre-gymnasium, gymnasium and post-gymnasium training in "forest counties," urban, and other counties

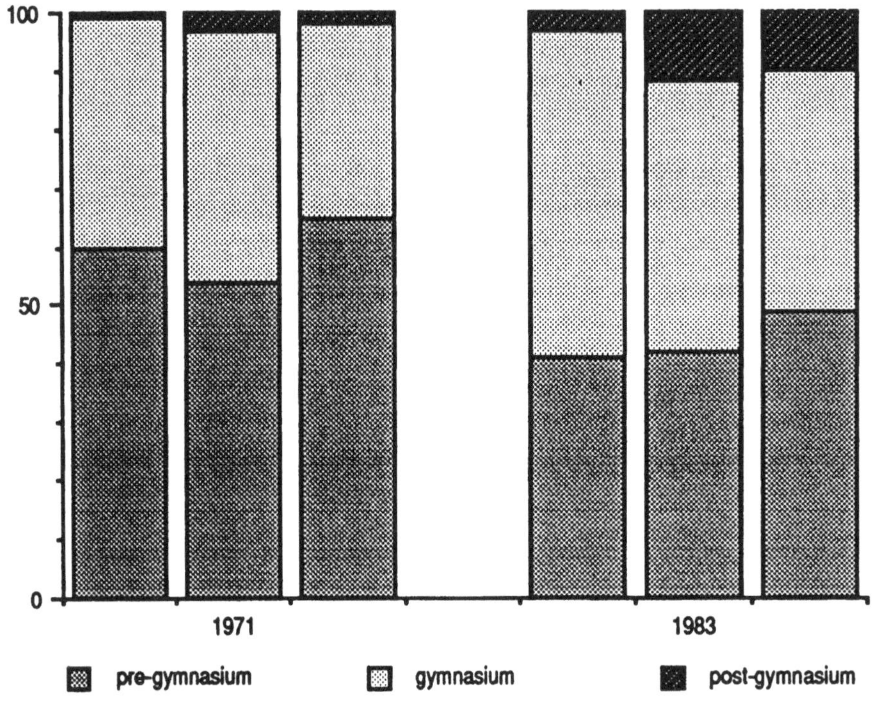

resents the typical peripheral region with an undifferentiated economy and a high unemployment rate. The latter county, in southern Sweden, is characterized by a differentiated population of small and medium-sized manufacturing firms and a dynamic economy.

The qualifications structures of small and medium-sized firms in the counties are different, although not in any clear-cut way to the disadvantage of the peripheral county (see Table 2).

Employees with "higher education" (qualified graduates in engineering, business economics, etc.) are significantly less represented in small firms in the peripheral county and to some degree also in medium sized firms. But vocational training is more common among employees in the firms of the peripheral region.

Table 2.: Employees with different types of formal training in small or medium sized industrial firms* in two Swedish regions 1985
(Percentage of employed in category of training)

Category of training	Small firms		Medium sized firms	
	Forest county	Other county	Forest county	Other county
Higher education	13	22	7	11
Vocational training	62	56	66	47
No vocational train	25	22	27	42
Number of firms	43	41	26	28

* Small firms = 5 - 19 and medium sized firms = 20 - 199 employees

The difference between the counties as far as employees with "higher education" is concerned affords no far-reaching conclusions. Within "industry," small and medium-sized manufacturing firms in the two regions have an almost identical share of personnel in this category. It would seem then that qualifications structures in different parts of Sweden vary less than would be expected if judged just by the rather different economic structures, migration, and unemployment histories of Swedish regions. It is probably correct to invoke public policy to "explain" this outcome. But it is hard to identify the decisive elements of the policy bundles involved.

The data on qualifications structures presented pertain to a period during which regional policy in Sweden, and in the rest of Europe, lost much of its initial formal precision. Administratively a decentralization process has evolved. Politically sector policies have made deep inroads into and contained regional policy proper. Empirically the effects of regional policy instruments have been hard to establish. Theoretically growth center and regional planning notions have lost momentum and key concepts such as "region" and "branch" of the economy are under review.

These developments warn against very ambitious attempts to correlate regional policy and developments of regional qualifications structures in the 1970s and 1980s.

4 Regional Policy and Qualifications Strategies in Sweden

Swedish regional policy commenced as an extension of selective labor-market policy. Hence, in the 1960s and early 1970s, while improvement of "human capital" was a much proclaimed national goal, its effects on regional qualifications structures were not necessarily positive. The idea was to qualify (laid off) labor for new jobs, irrespective of where, in a geographical sense, positions opened up. In effect, then, successful labor-market training schemes would not necessarily increase qualifications in areas targeted for regional policy measures. Rather, labor in such regions was mainly trained for the benefit of better developed areas, i.e., where the job openings were.

Until the late 1970s, regional policy in Sweden focused on moving capital and labor; presumably from developed to underdeveloped, from urban to rural areas, but, as measured by migration, with results that, at best, somewhat checked movements in the opposite direction (Elander 1978:108). The 1970s are also characterized by a state planning rather than a local mobilization conception of regional policy. Within the state planning concept administrative decentralization measures were introduced, however. This reflected developments to the effect that regional and not central state authorities had started to take charge of regional policy formation. But the available policy instruments changed only marginally.

In the Swedish political agenda, throughout the 1970s and 1980s the proclaimed goals for regional policy become more and more decoupled from labor-market policy and oriented towards creating jobs rather than reducing unemployment. Regional policy aims change first from an emphasis on state regional planning to an intraregional focus - from the central state as arbiter to regional agencies as coordinators of the regional policy instruments of the state. This re-orientation, effective from the late 1970s, meant more attention to regional competence and qualifications structures than before. But significantly it refers more to the agenda of some regional (and national) state agencies than to the regional policy programs they administer.

Recent developments have strengthened the "human capital" and qualifications focus even further. More and more, regional mobilization strategies and creative regions are being stressed. This, of course, again highlights the competence and R&D factors as determinants of regional development. But the part played by regional policy instruments to further the agenda could easily be misjudged. Regional state agencies, with a regional policy mandate, participate actively in the implementation structures of qualifications measures. The share of regional policy funds used (1986) for "soft investments" approximates, however, just some 10% of total regional policy investments (Nordic Council, Basprojektet 1987: 80f). The regional policy instruments proper are not adequately adjusted to such investments.

Two broad developments of significance for regional qualifications structures have been implemented. One is the expansion of the Swedish university system since the early

1970s. This national policy has had some "measurable" effects in northern Sweden. But indirectly, with results that are so far difficult to evaluate, the new universities/colleges have also become important as partners in a rapidly developing "new" policy area in Sweden: local industrial policy (LIP).[2] The universities/colleges belong to the institutional core of LIP, but several new resource centers are also affiliated with technical "gymnasia," and the formation of LIP is heavily influenced by the needs of municipalities and firms.

5 Regional Qualifications Structures and the University System

Until the mid 1960s university training in Sweden was concentrated in five locations in southern Sweden. North of Uppsala/Stockholm, Umeå University was established and enlarged during the late 1960s; regional development reasons were important for the localization and the effects on qualifications structures have been assessed and verified all around peripheral northern Sweden. The first technical university for research and higher education in the north was established early in the 1970s (Luleå). Later in the decade the regionalization of the Swedish university system started (1977/78). The regional development aspect of the new units was explicitly put. It was stressed that the new colleges should develop in close contact with the needs of local "industry" and provide opportunites for university training to new groups of individuals.

In the early 1980s universities/colleges were established in twenty locations across Sweden (including the seven "older" units). Technology transfer tasks were not consistently considered for the new colleges; nor, in the early 1980s, for the established universities. Nationally, technology transfer issues attained a more systematic treatment from 1983/84 on.[3] For the colleges to formally take on this task an addition to their mandate had to be made, since research had been defined as outside their authority (on the insistence of the universities). Government statements to this effect ensued in the following years.

Meanwhile, the colleges had become embedded in the local strategies of municipalities, firms, and regional state agencies. Several national boards responsible for technology and industrial development as well as education had engaged themselves in the "new technology" cause, frequently with the new colleges as the local end of programs. In 1985 the government asked four boards to make an inventory of ongoing technology transfer activities and to evaluate them. Concurrently the state county administrations were asked to document regional development potential and to help coordinate technology transfer measures in the regions. Reports and documentations have been published from 1987 on.[4] Evaluation activities are ongoing.

As the technology transfer agenda in the mid-1980s gained momentum in Sweden, the role and expectations of the new colleges (and some of the newer universities) were rather drastically enhanced. Colleges with no officially defined research programs were judged to have the potential of becoming regional assets in applied R&D in addition to fulfilling their original training mandate. The change in attitudes was not only a function of local endeavors but of some central aspirations as well; in this period the international competitiveness of Swedish technical training was put into question by several official reviews and a lack of interest in the field was discerned.

The fourteen new colleges established in the 1980s were small, and mainly provided teacher training. In terms of yearly student intake, the average number of slots in 1981 was some 450, half of this for teacher trainees. In 1986 the average number of slots had increased marginally, but the share of teacher trainees was just about one-third. Slots for technical training more than doubled from 1981 to 1986, and training in administration/ ecomomics/social work (AES sector) increased by 25%. Especially in the field of technical training, but to some extent also in the AES sector, the colleges helped rapidly to increase capacity in higher education (below the graduate engineer level). Between 1981 and 1986 their share of higher technical training increased from 14% to 22% (cf. SIND PM 1986: 8-10).

The average college is still small as far as number of yearly intake slots are concerned, but its contribution to the improvement (from the point of view of the "forest counties") of Swedish qualifications structures has been ascertained by interviews in firms and public agencies in northern Sweden. The impact of the colleges cannot be proved statistically, however. Studies indicate that the establishment of universities/colleges has significantly improved the degree to which high school students attend the types of courses offered in "their" region.

The technical university established in 1971 in Luleå in the most northern county in Sweden (Norrbotten) initially recruited as much as 80% of its students from the four northern "forest counties," with a more than "equal" share from Norrbotten. The proportion of "northerners" has decreased to some 60% lately, but the more telling figures refer to the whereabouts of the graduate engineers of the university. Of the 875 graduates leaving the university from the start, some 60% in 1986 worked in the four northern counties (Åberg 1988:216ff); a percentage that approximately corresponds to the recruitment share of these counties. A more detailed analysis of their job locations and work places indicates that Luleå graduates working in small firms are overproportionately represented in the most northern counties (70%) compared with those in large firms (50%). No comparable data for engineers leaving the new colleges are available. Indications of their involvement in the qualifications strategies of their regions may be gained from information on the volume of training courses and R&D activities commissioned by external actors (firms and public agencies); from 1981 to 1986 the volume of commissioned courses and R&D increased rapidly.

The market for commissioned types of university activities developed throughout the 1970s and early 1980s, and the formalization of these practises ensued by a government ordinance in 1986. The colleges take an overproportional share of the market; some 50% of total commissioned activities to the university system although their share of teachers/researchers is very much lower. Other indicators point in the same direction. In 1987, fees for commissioned courses, for example, on the average approximated 3% of total costs for basic training in the university system; in the colleges these fees averaged some 9%, however.[5]

The colleges' share of commissioned R&D is very low compared with the totals for the Swedish university system. Since, from the start, they were supposed to become only marginally involved in research activities, the interesting fact is not their share but that their commissioned R&D increased rapidly in volume during the 1980s; it more than tripled between 1981 and 1986. Not all of this development pertains to R&D related directly to local economic development. But at least in the colleges with comparatively larger budgets for commissioned R&D, the activities indicate that their competence is acknowledged by local firms and public agencies as well as by national research funds. Several of the colleges have organized R&D activities in "centers" sponsored together with firms, municipalities, and state agencies. By now most colleges, in one way or another, are expected, and not only by local/regional actors, to help provide access to networks of knowledge and skills that will accelerate the pace of local economic development.

6 Qualifications Policy and Technology Development Centers

The regionalization of the Swedish university system provided some twenty municipalities with units of higher education and some research facilities. The role of the new colleges may easily be exaggerated. They are, in many regions, the one readily visible institution of R&D activities. It is estimated, however, that R&D personell directly and indirectly (in resource centers, etc.) involved with the colleges amounts to some two hundred persons. In regions where colleges dominate higher education, R&D personnel in firms approximates between five and six thousand persons.[6] However, to public agencies in the regions, the colleges represent a more "controllable" focus of local qualifications strategies than private firms. If successful, the colleges may provide means to engage, in a more systematic fashion, resources from the private sector in regional qualifications efforts.

But while the colleges have started to form part of the implementation structures of local industrial policy (cf. Beckman 1987), most municipalities in Sweden rely upon other "centers" to improve qualifications structures. While 59 (21% of all Swedish) municipalities in 1988 were engaged in the activities of centers affiliated with universities/colleges, 133 (47%) were involved in "technology development centers" and some 230 (82%) were committed to institutionalizing closer links between municipal high schools and (mainly) local firms.[7] In fact, most Swedish municipalities try to establish themselves on the "new technology" agenda - in one way or another. With or without state support, in the 1980s, the municipalities have very markedly increased their efforts in the economic development field.

Until recently industrial policy was considered a state preserve in Sweden; meaning, according to state reviews, that municipalities lacked the overview, competence and resources to participate, if at all, as little more than executors of state policy in the field. During the 1980s several factors have activated Swedish municipalities in the industrial policy field. To discuss public policies for improving qualifications structures it is not necessary to go into details on what caused this activation. But it is important to describe, in brief, the changes it has meant for the formation of regional qualifications policy in Sweden.

Interestingly, in 1985, municipalities in the Swedish regions which for decades had been targeted for special regional policy measures by the state were also those that more frequently than others demanded a greater local say in matters of regional development (Hjern 1985:13-17). Most of them, especially in northern Sweden, had also started to organize administrative units for local industrial policy (LIP). This is a new policy field in Sweden which has developed rapidly. The profession of local economic development officers is now one of the fastest growing in Swedish local government (Hjern 1989a:22). Within LIP, making the local firms thrive and prosper is the first priority of (90% of) the municipalities. Having established firms relocate to the municipality is much lower on the list, mentioned with high priority by only some 50% of the municipalities (Hjern 1985:22-24). Good relations with local industry are now rated as much more important in Swedish municipalities than partnership with any other type of actor. One-third of the municipalities have developed programs for LIP in cooperation with local firms.

In sum, during the 1980s Swedish municipalities have not only taken on much greater responsibilites in industrial policy than they had before, but they have also, irrespective of political party rule, made local firms the focus of their strategies in a more consistent manner than before. However, at least in the "forest counties," their increasing professionalization has committed almost the same attention as before to the parts of their networks representing regional state administration units.

A decentralization process, not only within the state administration but also to the municipalities, is now evident in Swedish regional economic policy. It is a change towards local industrial policy dependent on the differing abilities of municipalities to engage themselves with (resourceful) firms and public agencies in the organization of local projects. Consequently, improvement of regional qualifications structures in Sweden to quite an extent has become the task of municipal mobilization strategies and "creative municipalities." Contrary to the myth about centralized Sweden, much of the recent surge to improve qualifications structures is coordinated from below.[8] State regional policy programs are important in the process. But, increasingly, so are resources from municipalities and private firms as well as from state agencies without regional policy mandates.

The establishment of "technology development centers" in the municipalities is not part of any one easily identifiable state policy. Centers affiliated with colleges have frequently had "seed money" from "project funds" of the State regional agency. This funding is limited to three years and to approximately one-third of the initial costs. Regional (county councils) and local government funding is generally required and contributions from industry are expected. A number of these "technology development centers" have had assistance for premises and equipment from regional policy funds. The "technology development centers" affiliated with colleges usually have an explicit technology transfer purpose. In most cases it is too early to assess if they will succeed. Early warning signals are given by reviewers, however.

The core of "technology development centers" which are not affiliated with universities/colleges consists of municipal technical high schools, sometimes jointly organized with significant firms in the local economy. Typically, such centers emanate from municipal attempts, made early in the 1980s, and sponsored experimentally by the National Board of Education, to upgrade courses of the upper secondary technical schools to the requirements of local industries. In some municipalities, with "high tech" industry establishments, these experiments in upgrading have also secured the backing of other national boards, e.g., the Swedish Board for Technical Development or the Swedish Industrial Development Board. Using "multipocket budgeting" techniques, several municipalities have extended their original aims of upgrading courses of technical schools and started to organize local "technology development centers" (see Hjern 1987).

7 Regional Qualifications Structures and Regional Policy

From the early 1960s to the late 1970s regional policy strategies in Sweden formally relied upon instruments controlled by the state, first with centralistic then with more decentralistic notions, which, however, all along have retained the model of the central government as the arbiter in matters of regional development. Dissatisfaction with results has led to several reviews to reprogram regional policy during the 1980s. Originally conceptualized for relocation of industries, well-placed observers claim that regional policy instruments still lag behind a practice moving towards local industrial policy.

Meanwhile, Swedish municipalities have started to rapidly qualify their own competence in matters of local economic development. This is happening at the same time as the technology training and transfer agenda gains momentum at all levels of Swedish politics and public administration. The increased involvement of local government in industrial policy formation partly ensues through the (uncoordinated) schemes of several national boards concerned with upgrading courses in technical schools to the needs of firms or with technical transfer experiments.[9]

In sum, as dissatisfaction with traditional regional policy instruments leads to reviews and while concepts such as qualifications of technical training and technology transfer find their way through the political, goal-setting machinery of the state, Swedish municipalities are rapidly establishing themselves as key actors in the formation of local industrial policy. For many of them, not only in those parts of Sweden officially defined as underdeveloped, an improvement of qualifications structures is not only an important policy aim but is considered a matter of survival. Consequently, all around Sweden, intensive local and regional efforts are being made to establish "technology development centers," often with the help of state agencies pushing the trendy new technology agenda.

Part of the technology agenda, in the mid 1980s, after some reconsideration of their original purpose, can be found in the new colleges. Most likely, the colleges and the "technology development centers" affiliated with them, in the near future will help the state to regain some control lost over the "new technology agenda." The spur of local activity, during a period of high performance in the Swedish economy, has left Sweden with too many public "technology centers".

The fate of many of these centers is at stake. The losers will be found in those municipalities that relied too heavily upon the experiments of the National Board of Education to upgrade courses in the upper secondary technical schools. Funding from the Board, which could be used for operating costs, will be cut. The upgraded courses are formally at the post-gymnasium level and consequently to be administered by the universities/colleges, and not by local government boards. Only larger municipalities will soon be able to position their technology development strategies within this new administrative set-up to create support for their centers. Regional policy arguments will not help centers of smaller and peripheral municipalities to attract funds in the future. By adding the "market shares" of smaller centers, the colleges, and the development centers affiliated with them, can expand and make their own R&D activities more viable. This development is supported by the central government. It needs to regain some control over regional qualifications structures after the very expansive local developments during the 1980s. The colleges will provide a means to that end.

Concentrating regional qualifications to colleges/universitites will not arrest the developments in Sweden away from traditional regional economic policy and towards local industrial policy. The rapid establishment of "technology centers" forms part of an overall trend of local mobilization among Swedish municipalities and decentralization of state responsibilities in the 1980s. In this process, shortcomings have become visible in the legal and administrative division of labor between Swedish central and local government with regard to public responsibilities for local economic development. The industrial policy monopoly of the state has withered and with it the role of state agencies as arbiters in matters of regional development. It is still not clear who will appear on the scene to formulate what will have to be a regional rather than a national industrial policy agenda.

Footnotes

1 This is measured as the percentage of employed who would have to change to another branch of industry or class of formal education in order to equalize branch or qualification structures across regions.

2 Some factors explaining developments from regional policy towards local industrial policy (LIP) are discussed by H.-J. Ewers 1987: 339ff. in the setting of the Federal Republic of Germany. In the Scandinavian setting the trend is discussed by Fredriksson (1984: 175ff.) Grahm (1988) and NordREFO (1982).

3 See UHÄ Report (1983: 9) Government Bill (1983/84: 7) and Government Committee Report SOU (1984: 74).

4 See SIND PM (1986: 7) and the summary report Teknikspridningsuppdraget (1987).

5 In the colleges the commissioned types of courses increased from some 2.5 mill. Swedish kronor in 1981 to 29 mill. kronor in 1986 and reached 45 mill. kronor in 1987. Compared with the total market for commissioned courses these totals represent an insignificant market share each year. But the point is that new colleges in less than ten years seem to have established an equal market share to the much more diferentiated and older universities.

6 The numbers are aproximations but are reliable as indication of the realtion between public and private sector R&D personnel in most regions outside the more urbanized areas in Sweden. See SIND PM (1986: 44-45).

7 Data from a survey of all Swedish municipalities expecting the three big cities in April 1988. Part of this material is published in Hjern (1989a).

8 A decentralization process to municipalities characterizes the implementation of several other public tasks as well. See Hjern (1989b).

9 In an assessment to "Regional policy as a distinctive field" in the Nordic countries, "the lag of the policy behind actual needs will become the recurring topic of the discussion" (Oscarsson 1988: 14). Much of the discussion in Sweden of the proper role of municipalities in regional policy, for example, concerns formal administrative and legal categories that lag considerably behind the knowledge of what their actual involvement and responsibilities are. See Hjern (1988).

References

Åberg, R. (1988)
Arbetslivets omvandling. In: Chans för Norrbotten. F. Snickars et al.(eds). CERUM: Umeå university.

Andersson, Å. (1988)
K-samhällets Framtid. Värnamo: Prisma.

Beckman. B. (1987)
Att bilda FoU-organ mm. ERU-rapport 51. Stockholm: Department of Industry.

Ekdahl, L. and B. Lindström (1985)
Arbete i storstad, ERU:46, Stockholm: Ministry of Industry.

Ekstedt, E. (1988)
Humankapital i brytningstid. Stockholm: Allmänna förlaget.

Elander, I. (1978)
Det nödvändiga och det önskvärda. Stockholm: Arkiv.

Eliasson, G. (1985)
The stability of economic organizational forms and the importance of human capital. Stockholm: IUI.

Ewers, H.-J. (1987)
Zur Dezentralisierung der Industriepolitik. In: Arbeitsplatzdynamik und Regionalentwicklung. M. Fritsch and C. Hull (eds). Berlin: Edition Sigma.

Fredriksson, C. (1984)
Från industribunden regionalpolitik till ortsbunden industripolitik. NordREFO 1984:1-2.

Grahm, L. (1988)
Att välja regionalpolitik. NordREFO: Nordstedts Tryckeri AB.

Hjern, B. (1985)
Kommunal näringspolitik - ansvar och kompetens. Stockholm: Swedish Association of Local Authorities.

Hjern, B. (1987)
Påbyggnadskurser som näringslivsanpassning. Umeå: ABH Företags- och kommuninfo AB (mimeo).

Hjern, B. (1988)
Centralstatlig avreglering för lokal näringslivsutveckling. NordREFO 1988: 4.

Hjern, B. (1989a)
Kommunal näringspolitik: professionalisering och demokratisering. Stockholm: Swedish Association of Local Authorities.

Hjern, B. (1989b)
Decentralized Sweden. In: Discretionary politics. D. Ashford (ed.), 1989 Yearbook: International Review of Comparative Public Policy.

National Labour Market Board (1988)
Labour Market Statistics.

Nordic Council, Regionalpolitiska basprojeket (1987)
Regional utveckling i Norden. Årsrapport 1987.

NordREFO (1982)
Lokal näringspolitik. Oslo: Universitetsförlaget.

Oscarsson, G. (1988)
Sammanfattande analys och förslag till fortsatt forskning.
In: NordREFO, Om regionalpolitiken som politikområde. Helsinki: Painokaari op.

SIND PM (1986)
Små och medelstora högskolor - regionala kompetenscentra?
Stockholm: National Swedish Board of Industry.

SOU (1984)
Regional utveckling och mellanregional utjämning.
Stockholm: Liber Allmänna Förlaget.

Swedish Central Board of Statistics (1986)
Survey 1986.

Teknikspridningsuppdraget (1987)
Teknik och kunskapsspridning - slutrapport. Stockholm: Liber Förlag.

UHÄ Report (1983)
Forsknings- och utvecklingsarbete vid regionala högskolor.
Stockholm: National Swedish Board of Universities and Colleges (UHÄ).

4.5 Universities and Local Economic Development

Kathrin Peters

Universities are increasingly considered as a distinctive resource for local economic development. However, how to harness this resource remains problematic and, as will be shown in this paper, requires solutions specific to each case. The diversity and complexity of this topic will be illustrated with the help of three case studies. These are:

1. The University of Warwick (United Kingdom)
2. The University of Texas at Austin (United States)
3. Cambridge University (United Kingdom)

The analysis of Warwick and Austin is based on an international comparative study of university/local economy links. This project, undertaken jointly by Segal-Quince-Wicksteed, researchers from Warwick University and American and Swedish teams, also included the University of Chalmers in Gothenburg (Sweden) as a third case. The information on Cambridge is based on our long-standing interest in the "Cambridge Phenomenon".

1 Warwick University

The University of Warwick is one of the younger UK Universities (founded in 1964). Even for UK standards, it is relatively small with about 6,000 students. Due to cuts in central government funding of the university sector, Warwick University (as many other higher educational institutions in the UK) was experiencing financial problems in the late 1970s and was forced to find alternative ways of raising finance. This process was facilitated by a strong central university management committed to academic excellence as well as to a meaningful role in the local economy, building on recognised strengths in advanced manufacturing, biotechnology and business studies.

Warwick University is located at the southern edge of the Coventry/Birmingham conurbation, an area traditionally dominated by vehicle and general engineering industry with recent experience of serious decline. Due to a number of factors (among them the two oil crises, the recession of 1982/83 and generally increased international competition), 53,000 jobs (46% of Coventry's workforce) were lost between 1974 and 1982. Since the late 1970s, the University of Warwick has become involved in a range of major local initiatives. The most important of these are:

1. Strategic arrangements with large companies. Amongst them are Austin Rover, Lucas Industries and Jaguar. University departments are involved with senior management on a range of issues including training, research, technology transfer and company strategy.
2. Modular advanced part-time management studies. Both the Engineering Faculty and the Business School have set up courses for employees of local companies to update and develop their technical and management skills. One of these, the Integrated Graduate Development Scheme, has recently been awarded a national prize for successful cooperation between industry and universities.
3. University departments are conducting both sponsored and collaborative research. One of these projects is the Advanced Technology Centre (ATC) sponsored by Austin Rover. The building of the ATC was provided by the University but Austin is paying rent for the accommodation. The intellectual property rights for work sponsored by Austin are vested with the company. The ATC is seen as an important step to improve just-in-time production methods and Austin is planning to make the research results available to its suppliers.
4. Small business teaching and advisory services organized by the Small Business Centre. The centre is an integral part of the School of Industrial and Business Studies and works in close collaboration with the academic staff of the School and the Institute for Management Research and Development. The centre's objectives are to engage in education, training, advisory services, consultancy and research in order to develop the professional skills of the management of smaller and expanding firms.
5. Science park including incubator facilities. The science park was set up as a limited company by four sponsoring partners (the University of Warwick,

Coventry City Council, West Midlands County Council and Warwickshire County Council); the park includes a 27,000 square feet venture centre which was constructed with the help of a £1.25m investment of Barclays Bank.

The central administration of the University, as distinct from departments, has played an important supporting role in many of the links. In a number of cases it has provided initial funding for initiatives which, especially with regard to the provision of accommodation of various kinds, has been a prerequisite for collaboration and also a demonstration to companies of the University's commitment to working with industry. Companies have also been impressed by the speed and flexibility with which the University has responded to their needs. However, there are strong indications that the major factor underpinning successful collaboration with industry is the early decision by the University to develop as a high quality academic institution. Its success is indicated by having departments with national and international reputations. The high standards achieved have been important in a number of ways. First, although companies considered proximity to the University useful, it was not a sufficient condition for establishing links. If they had not perceived the University as having top-class expertise, in terms both of R&D capability and producing a flow of highly qualified graduates and post-graduates, then initial contacts would not have been established and relationships with the University would not have developed to the extent that they have. Second, in the United Kingdom, funding of universities is becoming increasingly concentrated in centres of excellence. Those universities that achieve high standards are most favourably treated in grant applications enabling them to sustain and expand their activities with beneficial direct and indirect employment impacts in the local economy. In 1986, when the United Kingdom University Grants Committee first applied a new funding formula which included an assessment of research quality as one component, Warwick emerged as the most favourably treated of all UK universities. Third, and probably most importantly, universities may represent significant attractions to some companies seeking new locations. The high national profile and standing that Warwick has achieved represents a significant comparative advantage for the area in this respect; it undoubtedly also enhances the area's quality of life and hence its attractiveness to mobile individuals and projects. There are already signs that new types of industry are becoming established in the area, particularly in south Coventry and south Warwickshire, and while improvements in the transportation network have played a large part in this, the existence of a high quality, outward-looking university has also been important. The indications are that the University will come to play an increasingly significant role in the area's economic development infrastructure in the future.

2 The University of Texas at Austin

The University of Texas at Austin is part of the Texas-wide university system, located in the capital city. In the United States, the state government is responsible for the major proportion of funding for public universities. In Texas, state politics demand a justification on the basis of economic development for continuing financial support of the university system.

Texas' local economy has traditionally been dependent on primary sectors (agriculture and oil and gas). During the late 1970s and early 1980s, a need to diversify away from these sectors gained high priority in state politics. Facilitated by generous research funding by the state government, the University embarked upon a range of ambitious research programmes which contributed to developing the perception of Austin as a "technopolis", both within and outside the state of Texas, and subsequently led to the attraction of numerous high technology companies.

The most distinct of these projects is the Microelectronics and Computer Technology Corporation (MCC) which is a programme of precompetitive research financed by about 20 Information Technology (IT) companies at about $75m/year. Research is conducted in four different programme areas (semiconductor packaging, VLSI/CAD design, software technology and advanced computer architecture). By some observers, MCC is described as the American response to the Japanese fifth-generation computer programme. The decision process that led to the MCC locating in Austin was subject to fierce national competition. Originally, there had been 57 competing cities of which four had been shortlisted (besides Austin/Texas the other three were San Diego/California, Atlanta/Georgia and Raleigh-Durham/North Carolina). Austin won the competition by virtue of five factors:

1. University strengths. Academic excellence was the most important of all criteria and generous spending on computer sciences and electronics in the late 1970s and early 1980s had ensured that Austin scored highly here.
2. University commitment to the MCC. This commitment was evident both with respect to financial support as well as research collaboration with the MCC. The University provided the land and financed the MCC building which is leased to the MCC for a peppercorn rent. Research collaboration is

now organized in a variety of programmes and consultancy agreements. For instance, members of the University are retained to consult on how to facilitate the process of technology transfer from the MCC to the sponsoring companies.

3. Financial support from the private sector. The Texas state government succeeded in enlisting the private sector to support the selection process. The possibility of attracting the MCC to Austin was widely regarded as a great chance to boost the local economy and the private sector was more than pleased to comply. Altogether, the financial incentive package amounted to $62.5m. However, most of these were not direct financial incentives but contributions "in kind" or specified financial transfers such as the land and the building for the MCC, $20m worth of endowed chairs, financial support for graduate students, favourable lending terms for mortgages.
4. Combined effort. Overall, Texas' attempt to attract the MCC was characterised by the impressive cooperation of all parties concerned (the university, the public and the private sector); this related both to the financial incentive package as well as to political support, for instance an assurance by the state government to support the MCC against possible anti-trust legislation in the US Congress.
5. "Quality of life". Contrary to, for instance San Diego, Austin offered residential property of high quality at achievable prices. This was considered as very important because the MCC would depend on attracting a large number of graduates and researchers for whom the availability of achievable residential accommodation would be an important criterion.

There is little doubt that the MCC has had a major impact in enhancing the image of Austin as an R&D and high-technology economy. After the MCC had moved to Austin, various high technology firms located there or began major expansions in the area. This was due both to the symbolic importance of the MCC decision as well as to the fact that many MCC shareholders moved all or part of their R&D activities close to Austin in order to facilitate technology transfer (Rogers and Gibson 1988). Another prestigious research project, the Semiconductor Manufacturing and Technology Institute (Sematech) with an even more ambitious budget ($250m) was attracted to Austin in 1988.

The MCC also has had a major impact upon the University of Texas at Austin. By 1986, the computer science department was receiving three times as many graduate student applicants as it did before the MCC came to Austin (Rogers and Gibson 1988) and many undergraduate and graduate students work part-time at the MCC. There is a continuous flow of new opportunities emerging for interaction of academics and industry and new models for two-way technology transfer.

Although the incidence of spin-out companies from the MCC so far has been small, it can be expected that a research consortium of the size and quality of the MCC will generate technology based companies at some point in the future.

3 Cambridge University

Cambridge University is one of the United Kingdom's traditional universities with a reputation of international excellence in science, engineering and medicine. The University is organized in a highly devolved collegiate structure with the university's central authorities taking a benign hands-off approach to industrial linkage. Cambridge itself is a small, relatively isolated market town with little history of earlier industrialisation.

The "Cambridge Phenomenon" represents a growth of high-technology industry on a broad sectoral base in which the leading role is being played by "start-up" and the growth of small locally formed and independent businesses. The firms are mostly engaged in research/design/development or in high-value, low-volume production; what bulk production there is is sub-contracted elsewhere. This "clustering" of small high-technology firms has in turn attracted new operations of large companies to the area, both those engaged in advanced technology and in financial and business services. There has also been growth in the small-scale financial and business services sector, due to and in turn reinforcing growth of the small high-technology companies. These characteristics together add up to Cambridge being a high technology location of growing significance in the UK. Altogether, there are now over 600 high tech companies in the locality employing about 13% of the workforce which is double the national average (see Segal et al. 1985).

However, some of the factors that have helped to create the phenomenon in the first place could soon become constraints on its development. The small size of the town which has meant that it has been relatively easy to achieve a "critical mass" of high technology firms has also led to pressures on the local infrastructure, on traffic and other amenities. The limited industrial history, the fact that there has never been heavy industry, large plants and unionised labour, originally considered as an advantage because it allowed for a high degree of labour flexibility, now means that many companies face serious staff shortages and will have to move elsewhere if they want to expand into large-scale production.

4 Concluding Remarks

The three case studies have demonstrated that there are many different ways by which a university can have an impact upon its local economy. These range from support of local established industry (Warwick) to attraction of new, advanced technology projects (Austin) and spin-out of knowledge based companies (Cambridge).

Of the factors that influence the role a university can play in its local economy some are external, beyond the influence of the respective university. Amongst these are:

1. The industrial, corporate and spatial structure of the local economy. The Warwick model of close involvement in the technical upgrading of large local companies in traditional industries would be difficult to transfer to a place like Cambridge wich has only little industrial history
2. The availability of resources for research and special projects, etc. Although in some cases, the shortage of financial resources has led universities to open up links with local industry, at any one point in time the resource endowment has to be taken as given and will influence how attractive a university is to local industry.
3. Policies of local and central government. This relates to the funding of the university sector, including priorities in the research agenda and cooperation with the universities.

However, other factors are internal in the sense that they can be influenced by a university's administration, for example:

1. The culture of a university and its management structure and style. In the case of Cambridge, the "laid-back" attitude of the central administration and the decentralised collegiate structure have not imposed any constraints on individuals wanting to set up their own companies; this has led to numerous cases where academics remained attached to their faculties/departments and colleges while at the same time experimenting with a business idea (soft spin-outs). If the business took off and the individual decided that the life of an entrepeneur suited him better than that of an academic, he could leave the university at a later point. However, nothing would hinder him from continuing with this dual existence on a long term basis.

2. In the case of Warwick, a strong central administration poised to play a role in the local economy and alleviate the university's financial predicament actively supported links between university departments and industry, for instance by providing finance and resources in kind. Interestingly, so far Warwick has not produced many spin-out companies. This can be explained partly by the prevailing spirit of large scale training and research projects which might discourage individuals from pursuing their own commercial ideas.
3. Policies towards industrial linkage and specific mechanisms used. Both the priority attached to industrial links and the flexibility and speed with which universities react to approaches by industry as well as bringing them about, proactively influence the value of a university to local industry. This also includes issues such as how much time individuals are allowed for work with industry, how much income they are allowed to keep for themselves and how work with industry is acknowledged in staff assessments.
4. Fields and quality of research. Although to some extent dependent on the level of local or central government funding, universities are in a position to influence the direction of research and teaching, for instance by appointments.

The key conclusion emerging from the analysis of the three case studies is that there is no universally applicable model. Each university or polytechnic must develop its own approach. However, a common feature of all three case studies has been that academic excellence has been a precondition for successful linkage.

References

Rogers, E. M. and D. V. Gibson (1988)
M-form collaboration in a high-technology development decision: Selling Austin to the MCC. Los Angeles, California

Segal, N., R. Quince and W. Wicksteed (1985)
The Cambridge Phenomenon. The Growth of High Technology Industry in a University Town. Cambridge,

5 Indigenous Strategies and their Application in Regions with Differing Growth Dynamics

5.1 Advanced Technology Policies and Strategies in Developing Regions

Roberto Camagni and Roberta Rabellotti

1 Introduction

Productivity, continuing innovation and technological advances may be considered to be the driving forces behind regional economic development. An understanding of the laws by which they grow and expand through time and space is crucial for the possibility of a wise response of the regions to the different challenges they are facing; including take-off problems, sheltering of local infant industry, and restructuring or reconversion problems. We want to argue that the cumulative and centripetal dynamic of innovation diffusion is a spontaneous process (Section 2.1) and that the least-favoured regions must adopt new technologies in order to reduce the gap between themselves and developed regions (Section 2.2)

However if it is clear why they should adopt new technologies, it is not evident how. It is important, therefore, to analyse the technological choice, the instruments to adopt and the preference areas. In Section 3 we analyse the fundamental conditions for innovation processes and suggest the sectors and the areas in which it seems most convenient to operate. Finally, in Section 4, we take into consideration some useful instruments for a technology strategy: interface agencies, technology poles, and joint ventures involving local enterprises, national and foreign ones and the public administration.

2 The Need for a Regional Dimension to Community RTD Policies

Innovation and technological development are nowadays unanimously considered to be the "conditio sine qua non" for coping with ever-increasing international competition and ensuring economic growth. The growing gap in innovative attitudes between Europe and its strongest competitors, the U.S. and Japan, has recently led national and Community authorities to launch important programmes addressed at strengthening domestic R&D structures and firms' receptiveness to new technological and organizational models.

The same philosophy may well apply to regions and regional growth. The speed with which product and process innovation is adopted by local firms, the firms' ability to absorb new technological ideas and combine them with their specific needs, and their capacity for shifting resources from declining to growing products and market niches, are the main determinants of each region's economic performance (Camagni and Cappellin 1985; Molle 1984). Purely spontaneous processes of innovation and technology diffusion under laissez faire conditions show a spatial pattern that leaves peripheral regions at a great disadvantage. The features of this centripetal bias of new technologies will first be reviewed and then the reasons will be discussed in calling for a heavy commitment of local, national and Community policies in the direction of research and technology development (RTD) in the least-favoured regions (LTFRs).

2.1 The Centripetal Bias of New Technologies

2.1.1 The Cumulative Nature of Technological Development

First, innovations, with regard to both products and processes, and particularly "radical" ones, carry a strong probability of being implemented at the "centre" of the economic system, at the international level as well as at the interregional level. However, what really matters here is not so much this widely accepted "law" itself, but the cumulative nature of subsequent technological developments - after the pioneering phase.

At the level of the economic system, rapid incremental innovations follow the earlier ones along "natural trajectories", and different innovations coming from different areas may merge into "innovation systems" or "technoparadigms" (Freeman and Perez 1986).

At the territorial level, a host of cross-fertilisation processes, of synergies among research and production units, of spin-off effects through skilled labour mobility, take place within the limited boundaries of metropolitan areas or of specialised "valleys" and "corridors", mainly thanks to face-to-face contacts or easy access to relevant information. These facts and processes bring along very powerful polarisation effects in research activities, innovation and advanced industries, different in nature but similar in their spatial consequences to Myrdal's "backwash" and "cumulative causation" effects which explained the spatial concentration tendencies of the first part of this century.

Also at the level of the single firm, the previous cumulative effects play an overwhelming role: in fact, innovation and technical change often happen along a difficult learning process à la Arrow, which for the most part takes place within the firm through subsequent technical and organizational adjustments to perceived problems (Nelson and Winter 1982; Pavitt 1984). This process is particularly important (and tedious) in the case of new technologies and in particular in the case of advanced flexible automation systems, due to their "integrated" nature and the impossibility of an easy, "off-the-shelf" choice (Coriat 1981, Camagni 1986a): these difficulties explain the slow and spatially conservative adoption pattern of these technologies (Camagni 1986b).

These centripetal trends are definitely disadvantageous for LTFRs, not just, as it was believed some years ago, because of a general backwardness of the regional structure, but also for another more important reason, the pathological nature of most urbanisation processes and the critical state of the urban environment.

2.1.2 Obstacles to Innovation Diffusion

As do many other spatial processes, innovation and technological know-how almost automatically diffuse over the territory, through a host of formal or informal information channels. Epidemic models of innovation diffusion have long since presented a mathematical description of this process, taking into consideration neighbourhood effects, inter-city linkages and filtering-down processes along the urban hierarchy (Hagerstrand 1967; Pred 1977).

The limits of this approach to innovation diffusion lie in the fact that only contact probability and information exchange are considered to be relevant in the process; this probability being only an inverse function of distance on the three aforementioned channels (Camagni 1985). In fact, there are other important "preconditions" for innovation adoption that have to be taken into full consideration, especially when the spatial dimension is included in the picture, namely profitability and adjustment costs from the old to the new technology (or production).

A first category of these obstacles to innovation diffusion falls into what Mignolet (1982) calls the "accessibility to information". Some case studies on less-favoured regions in Europe, such as Sicily and Ireland, have shown that what is really missing in these areas is not just technological information, but marketing and commercial information as well (Camagni and Cappellin 1985).

A second category of obstacles derives from the characteristics of the local firms, the potential adopters, and it defines their "receptiveness" to innovation. An industrial tissue of small and medium-sized firms, highly specialised in the different production phases or in differentiated products of the same industry proves to be highly efficient in promoting very rapid imitation rates in process and organizational innovations ("systems areas", mainly located in the "Third Italy"; see Brusco 1982; Garofoli 1985). As well as the horizontal linkages among firms exploiting the same externalities or the same local know-how, we have to emphasize the fruitful vertical linkages along a "filière" (e.g. textile production and textile machine tools in Tuscany, shoemaking and machinery for leather shoes in Vigevano, near Pavia, Garofoli 1983; see Camagni and Cappellin 1985; ADEFI 1985).

By the same token, while an obligopolistic structure, made up of a few big firms, may be highly innovative within itself, it seldom provokes spin-off effects on the external environment, as the main information and service flows take place through internal channels instead. The Italian experience of a growth poles strategy for the South through creation of "white elephants" is sufficiently clear in this respect (SVIMEZ 1982; Camagni 1983).

A third category of obstacles to innovation diffusion, perhaps the most important, refers to those elements that help the firm overcome the "adjustments costs" from the old to the new technology. These elements characterise in a strict economic sense the external environment of the firm, and refer in particular to: (a) presence of production services and consulting agencies; (b) availability of skilled labour and management schools; (c) quality of the physical environment; (d) efficiency of the public administration; (e) industrial relations and; (f) social consensus on the general goals of the industrial society.

In a word, technical progress bears an intrinsic "relational" content: it needs, as Perroux put it, "le couplage d'un ensemble moteur et de structure d'acqueil" (Mignolet 1982; Aydalot 1986a). These environmental elements prevent diffusion of innovation and development processes from reaching some of the least-favoured regions, as some of the previous obstacles may prove crucial. Absence of R&D capacity and poor endowment of advanced services; big cities with broad, characteristics of backwardness and inefficiency, and small ones mainly addressed to agricultural trade; lack of entrepreneurial spirit and low level of local public administration; and conservative attitudes of trade unions in the face of scarce job alternatives for the labour force, are all negative elements for the diffusion of even imitative practices of technical change (Del Monte 1984; Cafiero 1985; Biondi 1985).

This assessment is corroborated by the results of the GREMI group on the innovative environments in Europe, synthesized by Aydalot's model of "rupture/filliation" (Aydalot 1986b, 1986c). If innovation (and development) always implies a break, a bifurcation with respect to past experience, it nevertheless finds its roots in the history of a local community, as it feeds on some pre-existing know-how synergy or tradition (Perrin 1986).

2.1.3 Long Waves of Innovation Diffusion and New Technoeconomic Paradigms

Long waves of spatial innovation diffusion occur in parallel with the imitation process of first innovations, with the expansion of systems of interrelated and incremental innovations departing from a few core ones, and with the consequent general development process. Starting from central areas, this process flows towards the periphery according to the specific conditions of each socioeconomic environment and the presence of favourable "preconditions" to its development.

Two other economic elements help this diffusion process: a general historical trend towards an increasing homogenisation of infrastructural endowment and educational conditions throughout the national space, and the evolution of each region's life-cycle. The first element is a structural, long-term one; the second plays a role in a shorter time span, and is responsible for the turning points in the development of success areas after long periods of continuing expansion (Camagni 1986b).

In years of a slowing down of the general innovation rate and productivity increase, as the 1970s certainly were, it is possible to discern a good performance in intermediate and even peripheral regions, where the "productivity to labour cost ratio" is more favourable than in central regions and the possibilities of entrepreneurial expansion and of "creative" imitation processes are present.

Is there any possibility that LTFRs may also benefit from the same process? The answer is only partially positive. Some areas, better endowed with modern urban services, R&D and local entrepreneurship may take off during the late phase of the cycle, as, for example, some areas in the Abruzzi, Molise, Puglia and Basilicata regions of Italy did, but many others, where the general "productivity to labour cost ratio" is comparatively unfavourable, may not.

But another element plays against the LTFRs' possibility of catching up: the appearance at long time intervals of new waves related to the emergence of new "technoeconomic paradigms". This fact may indeed cause a reversal of the previous trend towards decentralisation of economic activities, as the initial phase of the new wave once again requires "central" prerequisites: highly skilled labour and well-equipped metropolitan areas.

There are signs of this new trend (or at least of some important countertendencies) in the economies of some regions in Italy and France. Central and advanced regions like Ile de France and Lombardy are taking a lead once again in the 1980s after 15 years of slow growth, thanks to the development of all kinds of advanced activities, mainly in the tertiary sectors, linked to the new information technology paradigm (Aydalot 1986a; Gasparini et al. 1985). Once again LTFRs seem completely cut off from the natural spatial trajectories of the new paradigm.

2.1.4 The New Hierarchical Location Strategy of Large Firms

The hierarchical location strategy of the different internal functions, utilised by large multiplant firms is well known and widely accepted: headquarters and marketing functions are located in the core of big metropolitan areas, near financial and decision-making centres and close to commercial information networks; administrative functions, mainly white-collar ones, are in the outskirts of metropolitan areas, where mass transit systems are available and floorspace is cheaper; manufacturing and partial assembly functions are scattered throughout the world in developing areas, while final assembly and sophisticated productions remain more centralised; process engineering mainly follows the location of production plants; while research and sometimes development is increasingly located in pleasant environments, not far from well-structured metropolitan areas or advanced regions.

The start of a new long-term wave based on the microelectronic paradigm is likely to change this strategic picture in many respects in the next decades, as we may infer from some early signals. And the changes it is possible to foresee are destined to worsen, on the average, the situation for peripheral and marginal regions.

New technologies of flexible automation and computerized design and control of the production flow call for (a) an increasing functional integration; (b) new scope for reduction of working capital and consequently of needed floorspace; (c) a new general goal to be attained maximising economies of scope, or economies of joint production of differentiated products, instead of economies of scale; and (d) a new link between product and process innovation.

All these new elements (Camagni 1986b) are likely to bring along a new attitude in the spatial strategy of large and medium-sized firms. The need to integrate marketing, engineering and R&D functions and of new face-to-face contracts among people involved in these functions may lead to a baricentric locational choice, and therefore to a "central" one (Camagni and Rabellotti 1986).

2.1.5 The Centripetal Bias of Infrastructure Provision and R&D Public Programmes

As far as the effects of new information technology infrastructure on regional development are concerned, once again thorough reflection leads to a reversal of the conclusion arrived at upon first glance of a possible homogenisation of national space through the disappearance of the "distance" impediment. If it may be true that in the long term this condition could be approached, it is nevertheless certain that for some time to come infrastructural provision will first address the big advanced metropolitan areas, where demand density for new information networks is higher.

International networking policy brings additional support to this central bias, as it first links capital cities and major metropolitan areas; the case of Diane Euronet, an international overlay network among data bases, with one control centre in London and five exchange centres in Frankfurt, Paris, Rome, Zurich and London, is very clear in this regard.

Not only advanced infrastructural policy, but also policies addressed to the stimulus of technology and research, are likely to show a natural preference for advanced areas. Firms participating in the Esprit project, for example, are in fact mainly large firms, located in advanced areas as far as the headquarters and research facilities are concerned (Gillespie and Thwaites 1986).

In conclusion, RTD practices and advances depend strongly upon not only exogenous technological trends or firms' receptiveness to new ideas and new technological processes, but also upon the general "environmental conditions which are mainly under the responsibility of public decision-making centres" (CCE 1986). These environmental conditions prove to be particularly weak in the case of LTFRs, especially as far as new technologies are concerned.

2.2 The Need for RTD in the Development of LTFRs

The reasons why LTFRs do not easily adopt new technologies and compete in advanced production modes have been reviewed. In this section the reasons why they should do so are presented, while in the following sections some thoughts will be put forward about how they may be supported to do so.

2.2.1 The Equity Argument

A first general argument for why LTFRs need RTD involves interregional equity. As we have seen before, LTFRs are likely to lose in the interregional competition for modern production methods, and their relative level of well-being may be substantially reduced. The economic and social stability of the EEC and its member states may be endangered by growing economic polarisation. New technologies, if sufficiently mastered with the aid of public policy instruments, may show a still unproved potential for economic equity, increasing the diffusion of already existing manufacturing methods into local pre-existing, more traditional industries (Casey 1986).

2.2.2 The Efficiency Argument

The doubts about the social cost of a development strategy for LTFRs, centred on new technologies, may be overcome through consideration of the new features of regional economic development and the decreasing importance of an efficiency/equity trade-off. In fact, in an era of "indigenous", geographically rooted economic development, it is more and more true that national development results from the algebraic sum of each region's autonomous development, rather than being a quantum increase in wealth to be distributed among regions, as in zero-sum games. Capital, labour and also market capacity are no longer the constraining factors in interregional competition, while entrepreneurial ability in the utilisation of local, otherwise idle, resources plays the dominant part.

The revitalisation of local industries through new technologies and the increasing international competitiveness thereby acquired may prove able to increase national aggregate growth and performance (Vasquez 1984, 1985).

2.2.3 The Increasing International Competition Argument

The aforementioned arguments gain momentum if the general conditions of future economic growth are taken into account. In a period of slow growth, international competition is destined to increase substantially, and the need for a strengtheing of local industrial structures becomes crucial. Even if this implies a short-term reduction in employment levels, due to increased productivity, no alternative exists to a technological upgrading of local production structures. Failing this, the long-term consequences are going to be disastrous (SVIMEZ 1982).

In this regard, it may be useful to recall what David Ricardo said about technical progress 150 years ago in his famous chapter "on machinery" (1819/1981: 396-397):

> "The employment of machinery could never be safely discouraged in a State, for if a capital is not allowed to get the greatest net revenue that the use of machinery will afford here, it will be carried abroad, and this must be a much more serious descouragement to the demand for labour, than the most extensive employment of machinery ( ) By investing part of a capital in improved machinery, there would be a diminution in the progressive demand for labour; by exporting it to another country, the demand will be wholely annihilated."

In more modern terms, it may be affirmed that regional development disparities are in fact disparities in productivity and competitiveness.

2.2.4 The Continuing Innovation Argument

A further argument in favour of the introduction of RTD into LTFRs comes from a consideration of new technologies not just as process innovations, but more precisely as preconditions for continuing innovation and production flexibility, two crucial characteristics of modern industrial competition. Advanced but stable technologies, typical of the previous mechanical technologies paradigm have to be replaced by flexible technologies, which enable a change in product characteristics in almost "real" time and force processes to follow this "must" at low costs (Cafiero 1986).

2.2.5 The Product Quality and Market Argument

On a deeper microeconomic level, it is possible to affirm that LTFRs cannot escape from adopting new technologies to retain market consideration. As their local markets are generally narrow, by definition, LTFRs have to sell their products on international, advanced markets. These markets require sophisticated characteristics both in product (quality, design, novelty and variety) and delivery conditions (time reliability, production elasticity for peak or unexpected demand) that may hardly be achieved through traditional production technologies and organizational methods. The effort in this respect would be double for LTFRs, as they have to catch up in both production technologies (FMSs, CAD-CAM) and new managerial styles, an element in which also advanced European regions show many weak points with respect to Japanese firms (CCE 1986).

2.2.6 The Comparative Cost Advantage Argument

New production technologies are undoubtedly labour saving, as they substantially reduce and in a long-term perspective even annihilate the need for direct labour. In the past ten years, in fact, in big industrial firms direct labour costs as a share of total costs decreased from about 25% to 5%-10% (Ohmae 1985). As a consequence, they are likely to highly reduce the comparative disadvantage of stronger areas in manufacturing functions, as far as the labour cost factor is concerned. To continue to rely on the labour cost advantage means forcing a drastic reduction in per capita wages, a fact that may perhaps suit only really backward areas of the Third World (SVIMEZ 1982; Camagni 1984).

As a likely effect of these changes, LTFRs of the European Community may be crushed between advanced and backward areas. The only escape from this unhealthy position is the pursuit of a rapid technological upgrading of existing production methods and a convinced strategy of catching up with respect to more advanced areas through different forms of technology import, a necessary precondition for the possible subsequent "creation" of a local technology.

2.2.7 The Dependency Argument

Since Schumacher's seminal paper on "intermediate technologies" as a wise response to development problems of less-favoured areas (1974), a lively debate has taken place at the international level on the problems of advanced technology transfers and their "appropriateness" to development countries, with results that, as usual, have been interpreted and assessed differently.

The terms of the problem remain almost the same in both spatial environments. On the one side, we have the advocates of "indigenous", "soft", labour-intensive technologies, calling for new small firms and a spatially scattered, rural industrialisation; these technologies are claimed to be socially appropriate, self-reliant, manageable, pleasant and environmentally sound (Ghosh 1984). In the to date successful Third Italy experience, a "small is beautiful" slogan may in fact be highly appropriate for depicting real trends. On the other side, however, stand the advocates of a more advanced conception of "appropriate technologies", calling for an "appropriate design and adaptation" of best technologies to the production needs and "vocations industrielles" of the individual areas (Balcet and Colombo 1980) and relying on the flexibility and pervasiveness characteristics of microelectronics and biotechnologies.

The "blending" of best technology elements with traditional industry practices seems to be one of the new key strategies for a catching up process in LTFRs, a strategy which looks similar to many recent development stories: Japan and Italy during the 1960s were showing the highest deficits in their technology trade balance but were at the same time the countries with the highest, technology driven, development rates (Emmanuel 1982).

The wide possibilities of microelectronics application to traditional production, at a first stage, and to more advanced but still locally rooted productions in a second stage, suggest the strong potentialities of a different strategy with respect to the "intermediate technology", "pauperistic" and mainly "assistential" approach (Perez 1985).

3 "Vocations Productives", Regional Needs and RTD Strategies in the Case of LTFRs

3.1 Which Technologies for Which Purposes?

From the analysis presented in the first part of the present report, the cumulative and interactive character of technological advances is apparent. Demand and supply of innovation factors interplay and strengthen each other in the territory, giving rise to spatial concentrations of innovative poles. Advanced firms rapidly innovating in regard to both processes and products, enrich the environment with their technological and organizational know-how, while the general economic environment in its turn supports their activity and receives from their presence a fast-growing demand.

A process of this kind happens spontaneously in core areas, but needs time to develop. Its replication in LTFRs is by no means an easy process, especially if short-run effects are expected. What is certain in this respect, however, is that one-sided policies, addressed only to either element of the process, are likely to remain highly ineffective. The push comes from the general purpose, advanced infrastructure, as R&D facilities or information and telecommunication networks will generate development per se (O'Brien 1981); while the pull is from externally driven high-tech units which will never integrate into the local economic structure, unless the general environment secures the necessary services and modern infrastructure. This brings us to a first conclusion: a policy of RTD push should be carefully addressed to a potential demand already present "in nuce" in LTFRs.

Fig. 1: X efficiency of alternative techniques in the short and in the long run

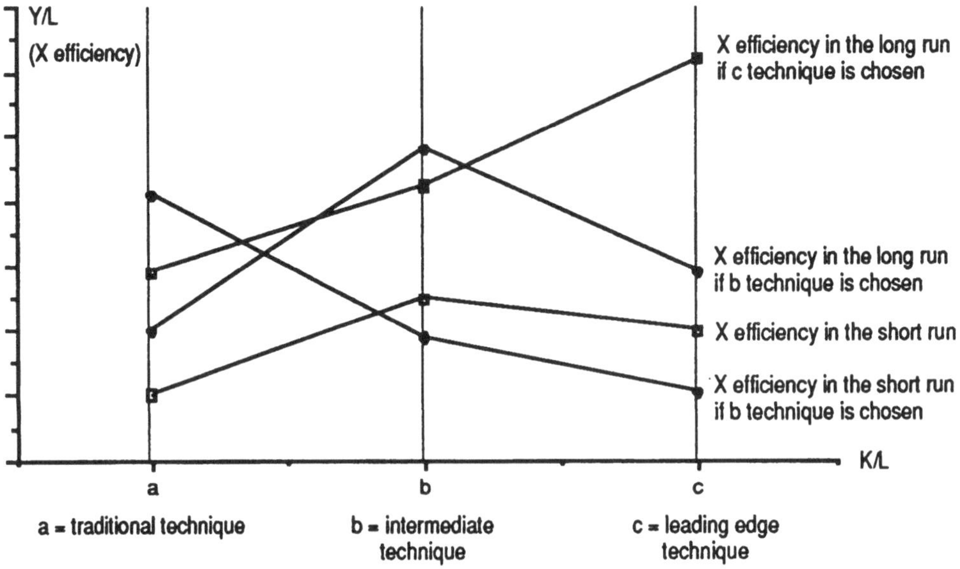

A second consideration regards the alternative between a strategy focused on "intermediate" or "leading edge" technologies. As it has already been shown, an intermediate technology strategy is advocated by many development institutions, like the OECD Development Centre or Unesco (Jequier 1976; Ghosh 1984) for less developed areas, and it has more recently been proposed with different emphasis also for developed countries.

Our point may be summarised as follows. An intermediate technology development may be useful for coping with severe unemployment problems, and in fact there are spontaneous mechanisms that work in this direction (the "black" economy is but one such example); however, surely this is suboptimal as a catching-up strategy. Especially in the medium and long run, the productivity gains it allows are not sufficient for standing up to international competition on advanced markets.

If we focus on the short run efficiency of alternative techniques, not the abstract efficiency of neoclassical schemes but the real "X efficiency" of these techniques when utilised in an unfavourable context, it may well appear that an "intermediate" choice (like choice b in Fig. 1) could show the best performance. But if we look at medium-term efficiency after a sufficient "learning by using" period, the answer will probably be different. Best practice techniques in fact are likely to enhance their own productivity and also the efficiency of more traditional ones, through spin-off effects on general education and infrastructure, while traditional or "appropriate" techniques will probably fail to do so (Stewart and James 1982).

In conclusion, the "appropriate" strategy for development of least-favoured regions within an advanced area (therefore benefiting from important externalities with respect to Third World countries) could be that of forcing leading-edge technology practices into local specialisation sectors.

3.2 Towards a "Creative Imitation" Strategy

Best technology practices may well be imported into less-favoured areas, and the fact is inescapable. Production of capital goods, even those that are less advanced, is a latecomer specialisation in the development process, and any development strategy, including one of import substitution or intermediate technology, would encounter the same problem. On the other hand, it is apparent that the importing of technology is a shortcut to catching up, as the Japanese and Italian experience as well as that of most NICs has shown (Soete 1981).

A policy of "excellence" and the push of "first innovation" would show a poor cost-benefit balance. Enclaves of advanced research could of course be realised within a backward environment, but with very low probability of external fertilisation of the local economy.

A better strategy could be that of rapid imitation of external productive patterns and of "creative adaptation of advanced technologies to perform specific tasks, close to the former "vocation" of the area. A better communication infrastructure may enhance the international penetration of local, even low-value added, production; small robotised systems may perform a host of simple but repetitive tasks in light production such as food processing, leather and clothing, plastics cutting, and even imitate, if their programme is enriched with random disturbances, handwork in artisan productions.

Due to their character of flexibility, reduced scale of minimum efficient adoption and easy or "friendly" use, many (but not all) information technologies fit well into a small business environment; others may be locally adapted to fit the same needs.

The biggest potential recipient of advanced technologies in less-favoured regions is the public administration, often the widest production sector. The computerisation of cadastre and health care, postal services and social security, represents at the same time a way of increasing sectoral productivity, social well-being and general efficiency. Computerised systems of agricultural production and computerisation of the tourist network (from transport and hotel reservation to itinerary planning and cultural assets inventory) would surely enhance the competitiveness of these important sectors.

Learning by using processes and selective efforts to adapt advanced technologies to local needs will in the medium run generate an increasing "technological mastery", necessary to eventually find some niches for indigenous advanced specialisations. A strategy of this kind requires selective action with respect to each regional reality. Therefore, a careful taxonomy of local development patterns and consequently of local RTD potentials is highly needed.

3.3 Internal Spatial Differentiation and Specialisation of Less-Developed Areas

3.3.1 Taxonomy of Spatial Patterns of Development

Less-favoured regions have to be considered as a diversified reality not only from an interregional, but also from an intraregional point of view. Within every region, different areas are characterised by specific paths of growth, according to their physical and geographic features, their historical development conditions and the kind of adopted policies.

From a mainly macroeconomic and sectoral point of view, some empirical inquiries made in the past decade have shown an increasing differentiation of less-developed areas in the EEC, and the emergence of interesting take-off episodes not just in the intermediate success "belts" but also (even if more rarely) in peripheral ones (Camagni and Cappellin 1980, 1981).

The existence of a diversified reality, of different development levels and paths involves the necessity (and the case) for a diversified intervention strategy. As stated before, this strategy should favour and strengthen the "vocations productives" of the different areas, exploiting the local internal know-how and the existence of a local economic "connective tissue". Therefore, it is an essential question to identify the different patterns of development, in order to properly address the intervention strategies in RTD in terms of both application fields and location.

In our view, it is possible to adopt a typology, taking into consideration, beyond aggregate indicators of development level and performance, also the degree of urbanisation, the type of industrial development and, possibly, general accessibility. The presence of a sufficiently structured urban hierarchy and of an endogenous and diffused pattern of industrialisation versus a concentrated or externally driven one, are important differentiating characteristics which are likely to deeply influence the receptiveness of local areas to advanced technologies.

We can classify different areas according to their historical patterns of development (or crises) by means of the following typology:

1. Development focused on urban centres (UD)
2. Diffused development in peripheral areas (PD)
3. Dependent development (recent branch-plants development; DD)
4. Crisis of (big) urban centres (UC)
5. Crisis of peripheral areas (PC)
6. Crisis of dependent growth-poles (DC)

As a theoretical exercise, this taxonomy may be applied in first approximation to the Italian Mezzogiorno experience, for 38 provinces (including assisted central Italian provinces). According to the 1986 SVIMEZ Report on the economic situation in the South, we can analyse the regional differences taking into consideration both static indicators, such as unemployment rate, per capita product and industrialisation rate, and dynamic ones, such as emigration quotient, recent increase of per capita product variation and rate of creation of new firms. The result of the SVIMEZ analysis is a classification of Mezzogiorno provinces according to their actual development status and their development trends. The SVIMEZ results may be supplemented by another classification of Southern provinces, based on a factor analysis emphasizing the degree of urbanization and tertiary specialization (Caniglia et al. 1985)

On this basis and in accordance with our suggested taxonomy, we identify three different general realities: a development situation, a crisis condition and an intermediate one (Table 1).

Within the progressive areas we can find, first, episodes of interesting urban development in medium-sized centres where an infant production services sector is taking off and a positive urban effect is strengthening the dispersed industrial development. Second, we can find wide belts of diffused industrial development in mainly rural areas. In these areas, industrial development is not confined only to larger urban centres, but seems to foreshadow a polycentric functional structure, branching out into the hinterland and involving both the small and medium-sized urban centres (Mattoscio 1984). With regard to the industrial structure, we find small firms, working in traditional sectors and characterised by local, spontaneous entrepreneurship. Third, we can find new development poles in advanced sectors where dependent development does not prevent the areas

Table 1: Taxonomy of assisted Southern Italian Provinces

Position \ Trend	Progress		Stationarity		Regress	
Favourable	Latina	UD			Siracusa	DC
	L'Aquila	UD-DD			Taranto	DC
	Pescara	UD-DD				
	Teramo	PD-DD				
	A. Piceno	PD				
	Frosinone	PD				
	Rieti	PD				
	Campobasso	PD-DD				
	Chieti	DD				
	Bari	UD				
Intermediate	Casserta	UD-DD	Catania	UC	Cagliari	DC-UC
	Avellino	PD	Napoli	UC-DC		
	Isernia	PD	Salerno	PC-UD		
			Potenza	PC-UD		
			Sassari	DC-UD		
			Brindisi	DC		
			Ragusa	DC		
			Foggia	PD		
			Matera	DD		
Unfavourable	Lecce	UD-PD	Messina	UC	Palermo	UC
			Benvenuto	PC	Trapani	PC
			Consenza	PC-UD	Nuoro	PC
					Oristano	PC
					Caltanissetta	PC
					Agrigento	PC
					R. Calabria	PC
					Catanzaro	PC
					Enna	DC

UD = Development centred on urban centres
PD = Diffused development in peripheral areas
DD = Dependent development (recent branch-plants development)
UC = Crisis of urban cemntres
PC = Crisis of internal / marginal (peripheral) areas
DC = Crisis of ancient growth-poles

from experiencing a rapid growth in both employment and per capita income, even if in some cases a restructuring process is now under way.

Taking into consideration the crises realities, we can identify three main patterns; the economic and social crisis of the largest metropolitan areas; the growth pole crisis, when the crisis of a sector causes the crisis of a whole area; and the crisis of marginal areas, where peripherality represents an obstacle to development.

3.3.2 A Selective RTD Strategy for Each Development Pattern

On the basis of the preceding taxonomy of local development patterns it is possible to detect the main lines for a selective RTD strategy in LTFRs.

First of all, we must recognize that a RTD policy does not exhaust the possible spectrum of development interventions in less-favoured areas, and that it shows some peculiarities. In fact, within the traditional framework we are exposing, it requires some basic local preconditions for its effectiveness, which highly restrict the spatial context of its applicability. In particular, it requires a sufficiently structured and dynamic urban or industrial environment, in order to avoid two different possible negative effects: the lack of supply of advanced production factors (and consequent import of skilled personnel) and the absence of local demand of advanced facilities (with consequent failure of the intervention effort).

It is important to emphasise that a criterion based on performance can obviously contrast with one based on structural features, and, therefore, to penalise the big metropolitan areas, where innovation processes should be located according to the theory. However, this is true only for advanced areas and not necessarily for areas in which urban growth is not due to economic development.

As a consequence, an RTD policy should address those areas which already present some characteristics of modernisation; in terms of our taxonomy in Table 1, this means the provinces belonging to the upper-left triangle: advanced interventions in the other part of the table should be carefully assessed. In these latter regions, priority should be given to policies addressed to build (or rebuild) preconditions for growth: transport infrastructures and social overhead capital provision in internal areas, urban renewal and social policies in big cities, and economic reconversion or diversification in distressed "growth pole" areas.

These reflections provide us with a first, locational guide-line for intervention strategy. A second guideline concerns the character of the interventions, and this is linked to the typology of the local development pattern: diffused industrial development, dependent and concentrated development and urban-centred development.

In the first case, the goal of RTD strategy should be the increased competitiveness of the local specialisation industry, through adoption (and "creative adaptations) of advanced information technologies and flexible manufacturing processes. From a purely technological point of view, the specific typology of applied and development research facilities should depend on the specific local "vocations industrielles". But it should be firmly kept in mind that technological transfer to small and medium-sized firms is hardly just a problem of technological information (and financing), but rather a complex organizational and strategic planning problem. Therefore, a consultancy structure in these latter fields should work in strict contact with the research and engineering units in each territorial context. By the same token, the usefulness of advanced facilities in the field of specialised commercial and market information would prove particularly useful and appreciated. The "blending" not just of advanced technologies and more traditional processes, but also of advanced managerial practices (e.g. in the programming of the internal and external flow of inputs and outputs) with more traditional ones, especially in the relationships with the labour force, have given interesting results in other semi-advanced contexts, and is likely to constitute a new and perhaps "appropriate" model also for many LTFRs.

In the second case, the dependent development is based on the recent establishment of big branch plants of external firms, both in high-tech or in more traditional sectors. The case shows up for a triangular agreement between the firms, representatives of the local community and the central public agency for RTD intervention, addressed to an upgrading of the functions attributed to the branch plant, with the parallel support of external, publicly-financed facilities, working in the fields of advanced professional and managerial education, development (or sometimes even basic) research and communication infrastructure provision.

As in the previous case, it would be useful for us to try to push these suggestions further in the sense of dictating specific fields for intervention. In fact, these indications are part of the project design phase, which has to take place with the participation of local interest groups, the true final users (and managers) of the projected facilities.

In the last case, the one where some sufficiently equipped and socially healthy urban areas have sprung up, the intervention possibilities for RTD could be enriched through the use of many general-purpose policy instruments, not addressed specifically to some sectors but potentially to a variety of them. Examples of such instruments include: technical advances in public services management; provision of properly infrastructured floorspace for small and medium-sized firms, endowed with advanced, centrally man-

Table 2: Intervention strategies in LTFRs

Intervention Strategies in LTFRs	RTD Development policies	Advanced projects in conjunction with local universities and research centres	UD: urban centered development
		Upgrading of local branch plants functions; joint ventures in advanced projects	DD: recent branchg plants establishment
		Strengthening of local "vocations productives" through blending of traditional and advanced techniques	PD: diffused development and "Systems areas"
	Traditional Development Policies	Urban renewal; re-creation of a pleasant physical and social environment; Job-creation	UC: urban crisis
		Reconversion, restructuring and diversification policies	DC: crisis of first generation growth poles
		Accessibility and infrastructure policies; attraction of external firms	PC: desertification of internal and peripheral areas

aged, services, along the French tradition of "pepinières d'entreprises" (or "incubators"); product testing equipment managed by public research centres or engineering universities; and research centres in some particularly important fields of production for the local area. All these undertakings and many others, possibly organized in a spatially concentrated way in urban technology poles, in strict relationship with existing public research units and local universities in science, technology and business administration fields, could be usefully launched in these areas (Table 2).

4 Mechanisms for RTD Transfers and their Effectiveness

Traditionally the public strategies in the area of innovation directly involve the three components of the innovation process: technology (in the form of subsidies to basic and applied research programmes), firm (general investment subsidies) and environment (professional and university training programmes). However, these general interventions have often shown a limited economic impact, as they are not integrated into a coherent general strategy: the results of many important research programmes have not been exploited by firms, the availability of new technologies not always followed by a true capability to use them, while training programmes have often shown weak linkages with labour market needs. Therefore, particularly in the case of least-favoured regions, it would be better to address innovation policies to the areas where the three components should interact and where useful potential synergies may be created.

4.1 In Search of Synergies

4.1.1 The Environment Technology Integration Area

In least-favoured areas, intervention should be addressed to bringing the training and schools system closer to technology and science. From this point of view, the universities should activate degree courses in applied information science and electronics engineering, in order to give an up-to-date training to students; but above all they should emphasize their non-didactic functions in order to become a "creative" and attractive environment for basic and applied research centres and to supply advanced services to firms.

4.1.2 The Technology Firm Integration Area

In this area the possible interventions are: (a) financial subsidies to "intramuros" research, prototype production and initial commercialisation of new products, and (b) the creation of interface agencies, connecting firms with the research world. Typically the systematic search for (and evaluation of) new products and technologies applicable to the needs of a particular company calls for large company resources. Interface agencies assist firms, particularly small and medium-sized ones, to put advanced product innovation and technology transfer within their practical reach. Their role is to assist clients to define technological opportunities for expansion of their product range or diversification; to identify licenses, agents or buyers for client's technologies, and to assist them in the negotiations. In other words, the interface agency working as a catalyst of synergies, connects supply and demand of technology, not only providing technological information but also exciting a qualified demand and inducing technological needs in local firms.

4.1.2 The Firm Environment Integration Area

The firm environment integration area involves the availability of services to the firm for assistance, not only in the technological field, but also in the evaluation phase and in the internal adaptation process to new technologies. Organization and strategic planning consultancies are fundamental elements for a wise and effective use of new technologies: usually one of the main causes of the rate of failures among small firms engaging in new processes is simply lack of a broad range of management knowledge.

4.2 Technology Poles: the Technology Environment Firm Integration Area

The area which results from the overlapping of the three fundamental components of the process of innovation diffusion (the technological system, the environment and the firm) concentrates all preconditions for a rapid adoption of innovations.

In the technology poles the synergies are usually maximised: nevertheless, it may happen that this interaction could never reach a sufficient momentum, due to the weakness or simply non-existence of some subsystems.

Therefore, in the case of least-favoured regions, the necessary synergies have to be found not really on a local scale, but rather on a national and international scale, attracting the innovative efforts of different subjects and acting as a catalyst to external/internal potentials. Second, as already stated in the previous sections, the general involvement of the research activities should not be on frontier subjects, but on applied matters and particularly on the applicability of advanced technologies by the local industries. Joint ventures and technological cooperation agreements could therefore be promoted between external firms, both producers and utilisers of advanced technologies, on the one hand, and local research institutions and local firms, on the other.

The general goal should in fact be to generate a self-propelling development with significant spread effects on the local economic system, and not to build "ivory towers" or even "empty boxes" attracting only friendly public orders.

4.3 Joint Ventures and Technological Ccooperation Between External Firms and Local Ones

In least-favoured regions it may often happen that local know-how and entrepreneurship are insufficient to promote a self-sustained growth, even if stimulated through appropriate instruments of industrial and regional policy. Therefore, the appeal to external energies may prove necessary. On the other hand, high-tech firms have traditionally chosen least-favoured regions as locations for purely manufacturing activities, characterised by a high intensity of unskilled labour and the absence of advanced functions and of decision-making responsibility (Del Monte 1980b, 1982a). Therefore, intermediate strategies between the two extremes of autarchy and dependency have been explored, both by "donor" and "host" regions. This intermediate strategy has taken the shape of "new forms" of international and interregional investments, allowing both a reduction of the risk and the financial involvement of the investing firm and the growing responsibility of the local "acceptor" milieu (Oman 1980).

These new forms of interregional relationships allow the utilisation of the existing international know-how in science and technology in order to promote indigenous development, through different ways of economic cooperation, such as exports of complete industrial projects, joint ventures, or licensing (Momigliano and Balcet 1983).

4.4 Public Role in RTD Strategies

The public administration (P.A.) can assume a twofold role in RTD policies. The P.A. works in fact as a primary demand source and as a supplier of advanced services in communication, social security, health, education and environmental quality-control fields.

The demand may be crucial for the pace and the direction of innovation in the private sector: a public qualified demand, addressed to guarantee a continuous and stable demand flow of advanced devices and services, highly limits the entry risk for local producers and creates the necessary shelter for the endogenous "infant industry". On the supply side, the need for strong interventions in professional and managerial education is evident. But something more could be achieved through an information policy on technological opportunities and the ways by which they may meet the needs of the local production fabric. A long-term scheme for RTD upgrading in LTFRs should be based not only on a thorough analysis of such needs but also on the promotion itself of qualified needs and of a local demand for advanced technologies.

5 Conclusions

The main conclusions of this paper may be summarized as follows:

1. Technological advances, continuing innovation and productivity gains may be considered the driving forces being regional economic development.
2. Spontaneous processes of technology and innovation diffusion under laissez faire conditions show a spatial pattern that is highly disadvantageous for peripheral and less-advanced regions, especially when a new long wave based on a revolutionary technological paradigm, is taking off. The highly cumulative character of subsequent incremental innovations, and the strong obstacles to innovation diffusion posed by difficult accessibility, low factor skills and poor availability of entrepreneurial capability work as powerful accelerators of interregional economic disequilibria.
3. New flexible automation processes and new "information technologies" are likely to show the same centripetal spatial effects. On the other hand, the involvement of LTFRs in advanced technologies is to be considered a must, if the gap between advanced and backward regions is not to widen in the next decades.
4. The mastering of modern technologies is an inescapable task for developing regions. The real problem is how these technologies could be adopted, for which purposes and following which strategy. The "blending" of new techniques and organizational methods with local and traditional methods may give rise to new comparative advantages for developing areas, at least in an early stage; afterwards, successful adoptions in some leading sectors may have imitation effects on other sectors and produce favourable fall-out effects on many aspects of local economic life: training effects on the labour force, demonstration effects on entrepreneurship, and productivity effects on other sectors through backward and forward interindustry linkages.
5. The implementation of this general strategy poses many problems, in terms of sectoral and locational focusing. In fact an RTD intervention requires some basic local preconditions for its effectiveness, which highly restrict the

economic and spatial context of its applicability. In particular, it requires a sufficiently structured and dynamic urban or industrial environment, in order to avoid two different possible negative effects: the lack of supply of skilled production factors and the absence of a local demand for advanced facilities. The intervention should therefore be addressed to areas which already present some minimal modernisation characteristics and a proper urban (social and physical) environment. On the contrary, in the case where these requirements are not met, more traditional intervention policies should be envisaged; such as accessibility and basic infrastructure policies, urban renewal, attraction of external firms, and job-creation policies.

References

ADEFI (1985)
L'analyse de filiere. Economica, Paris.

Aydalot P. (1986a)
Les technologies nouvelles et les formes actuelles de la division spatiale du travail. C3E, Cahiers 47, mars, Université des Paris 1.

Aydalot P. (ed.) (1986b)
Milieux innovateurs en Europe. GREMI, Paris.

Aydalot P. (1986c)
Trajectories technologiques et modeles regionaux d'innovation.
Paper presented at the XXV Conference of the ASRDLF, Paris, September also in: Aydalot P. and Keeble, D. (eds.), (1988): High technology industry and innovative environments. Francis Pinter, London.

Aydalot P. and D. Keeble, (eds.) (1988)
High technology industry and innovative environments. Francis Pinter, London.

Balcet G. and V. Colombo (1980)
La peranza technologica. Etas Libri Milan.

Bianchi G. and I. Magnani (eds.) (1985)
Sviluppo multiregionale: teorie, metodi, problemi. Franco Angeli, Milan.

Biondi, G. (1985)
Innovazione ed assetto territoriale dell'industria meridionale.
Mezzogiorno d'Europa 3.

Brusco, S. (1982)
The emilian model: productive decentralisation and social integration.
Cambridge Journal of Economics 6.

Cafiero, S. (1985)
Il Mezzogiorno tra transizione technologica dell'intervento straordinario. Delta 17.

Cafiero, S. (1986)
I termini attuali del problema dell'industrializzazione del Mezzogiorno.
Studi SVIMEZ, 2.

Camagni, R. (1983)
Italy. In: Klaassen, L. H. and W. T. Molle (eds.): Industrial mobility and migration in the European Community. Gower, London.

Camagni, R. (1984)
Il robot italiano, Milano. Edizioni Il Sole 24 Ore.

Camagni, R. (1985)
Spatial diffusion of pervasive process innovation.
Papers and proceedings of the regional science association, vol 58.

Camagni, R. (1986a)
The flexible automation trajectory: the Italian Case. Paper presented at the international conference on "Innovation Diffusion". Venice, March 1986.
Forthcoming in: David P. and G. Dosi (eds.): Innovation and the diffusion of technology. Oxford University Press.

Camagni, R. (1986b)
Functional integration and locational shifts in the new technology industry.
Paper presented at the GREMI Seminar on "Technologies nouvelles et développement régional", Paris, September 1-3. Also in: Aydalot Ph. and D. Keeble, (eds.) (1988) High technology industry and innovative environments. Francis Pinter, London.

Camagni, R. and R. Cappellin (1980)
Struttura economica regionale e integrazione economica europea.
Economia e politica industriale, n. 6.

Camagni, R. and R. Cappellin (1981)
Policies for full employment and more efficient utilisation of resources, and new trends in European regional development.
Lo spettatore internazionale, Quarterly of IAI.

Camagni, R., R. Cappellin and G. Garofoli (eds.) (1984)
Cambiamento tecnologico e diffusione territoriale. F. Angeli, Milano.

Camagni, R. and R. Cappellin (1985)
Sectoral productivity and regional policy. Commission of the European Community.
Document, 92-825-5535-6, Bruxelles.

Camagni, R. and R. Rabellotti (1986)
Innovation and territory: the Milan high-tech and innovation field.
In: Aydalot, Ph. (ed.)(1986b): Milieux innovateurs en Europe. GREMI, Paris.

Caniglia, P. and L. Cavola, M. D'Antonio, F. De Felice and L. Limoncelli (1985)
Una radiografia del nouvo Mezzogiorno. Le trasformazioni intervenute nella società e nell'economia meridionale nel periodoItaliana di Scienze Regionali,
Genova; October 23-25.

Casey, T. (1986)
Diffusion of advanced manufacturing technologies into metals and engineering firms in the less developped regions of the European Community: issues in education and training from Irish case studies. Paper presented at the Convegno CEE-FAST on "Nuovi sistemi di produzione. Torino, July 2-4.

Commission des Communautés Europeenes (1986)
L'amélioration de la compétitivité et des structures industrielles de la communauté.
Communication de la Commission au Conseil, Brussels, February 25.

Coriat, B. (1981)
Robot et automates dans l'industrie de serie. Esquisse d'une economie de la robotique d'atelier. In: ADEFT, Les mutations technologiques. Economica, Paris.

David, P. and G. Dosi (eds.) (forthcoming)
Innovation and the diffusion of technology. Oxford University Press, Oxford.

Del Monte, A. (1980a)
Dipendenza, proprietà e struttura organizzativa delle imprese:
il caso dell'industria elettronica campana. L'Industria.

Del Monte, A. (1980b)
Settori maturi e settori innovativi nell'industria meridionale.
Quaderni sardi di economia.

Del Monte, A. (1982a)
Decentramento internazionale e decentramento produttivo. Loescher, Torino .

Del Monte, A. (1984)
Politica regionale, progresso tecnico e sviluppo nel Mezzogiorno. In: Camagni R., Cappellin R., Garofoli G. (eds.)(1984) Cambiamento tecnologico e diffusione territoriale. F. Angeli, Milano.

Del Monte, A. (1982b)
Le commesse pubbliche come strumento di politica industriale e regionale.
L'Industria, n.4.

Emmanuel, A. (1982)
Technologie approprié ou technologie sous-developpée. PUF, Paris.

Freeman, C. and C. Perez (1986)
Innovazione, diffusione e nouvi modelli tecno-economic. L'impresa.

Garofoli, G. (1983)
Collocazione internazionale e competitività del settore della macchine per calzature, cuoio, pelli. Assomac, Vigevano.

Garofoli, G. (1985)
Sviluppo multiregionale e sviluppo industriale. In: Bianchi G. Magnani I. (eds.) (1985) Sviluppo multiregionale: teorie, metodi, problemi. Franco Angeli, Milan.

Gasparini, I., R. Camagni and E. Marelli (1985)
La struttura economica: ruolo e capacità dinamiche di risposta dell'area metropolitana milanes. Rapporto di Sintesi, Progetto Milano, Milan.

Ghosh, P. K. (ed.) (1984)
Appropriate technology in third world development. Greenwood Press, Westport.

Gillespie, A. and A. Thwaites (1986)
Strategies to encourage the development of new information technology production in Europe's less favoured regions. Paper presented at the symposium on "Information and Regional Development". Delphi: April.

Hagerstrand, T. (1967)
Innovation diffusion as a spatial process. University of Chicago Press, Chicago.

Klaassen, L. and W. Molle (eds.) (1983)
Industrial mobility and migration in the European Community. Gower, London.

Jequier, N. (1976)
Appropriate technology: problems and promises.
Development Centre of the OECD, Paris.

Mattoscio, N. (1984)
Modelli interpretativi, struttura industriale e politiche d'intervento regionali: il caso dell'èconomia abruzzese. Rassegna Economica, 3.

Mignolet, M. (1982)
Estimation des disparites regionales dans l'access à l'innovation. Revue d'economie regionale et urbaine, 2.

Molle, W. (1984)
Potenziali regionali di innovazione nella comunità europea. In: Camagni R., Cappellin R., Garofoli G. (eds.), (1984) Cambiamento tecnologico e diffusione territoriale. F. Angeli, Milano.

Momigliano, F. and G. Balcet (1983)
Nuove tendenze nei processi di internazionalizzazione: recenti sviluppi del dibattito sulla differenziazione dei modelli di comportamento delle imprese multinazionali e su vecchie e "nuove forme" di coinvolgimento estero delle imprese. Economia e politica industriale, 40.

Nelson, R. and S. Winter (1982)
An evolutionary theory of economic change.
Harvard University Press, Cambridge Mass..

O'Brian, R. (1981)
Specialised information and global interdependence. In: Seers D. (ed.), (1981): Depandancy: a critical reassessment. Frances Pinter, London.

Ohmae, K. (1985)
Triad power. Free Press, New York.

Oman, C. (1980)
The new forms of investment in developing countries, the state of the art. Working Paper, Paris.

Pavitt, K. (1984)
Sectoral patterns of technical change: towards a taxonomy and a theory. Research Policy.

Perez, C. (1985)
Microelectronic, long waves and structural change: a world case. In: World Development, 13.

Perrin, J.C. (1986)
Les synergies locales: elements de theorie et d'analyse. Paper presented at the XXV Congress of the ASRDLF. Paris, September.

Pred, A. (1977)
City systems in advanced economies. Hutchinson, London.

Ricardo, D. (1819/1981)
On the principles of political economy and taxation. Edited by Piero Sraffa. Cambridge University Press, Cambridge.

Schumacher, E. F. (1974)
Small is beautiful. Abacus, London.

Seers, D. (ed.) (1981)
Depandancy : a critical reassessment. Frances Pinter, London.

Soete, L. (1981)
Technological dependency: a critical view. In: Seers D. (ed.): Depandancy: a critical reassessment. Frances Pinter, London.

Stewart, F. and J. James (1982)
The economics of new technology in developing countries. Frances Pinter, London.

SVIMEZ (1982)
Technological progress and development prospects in the Mezzogiorno in the 80s. Fast occasional papers 48, Brussels; March.

SVIMEZ (1986)
Rapporto sull'economia del Mezzogiorno. Roma.

Vasquez Barquero, A. (1984)
Los programmas territoriales, nuevo enfoque para la politica regional de Espana. Paper presented at the symposium on "Ristrutturazioni produttive e politiche di intervento nell'industria". Ancona, July 6-7.

Vasquez Barquero, A. (1985)
Regional policy in times of crisis: the Spanish case. Paper presented at the European Congress of the RSA. Budapest, August 1985.

5.2 Indigenous Development Strategies in Peripheral Regions: The Example of Ireland

Gerry Sweeney

The real rate of unemployment in Ireland lies somewhere around 29%. Of a population of 3.541m, 1.088m were at work in 1987. The unemployment figure for August 1988 was 242,000 representing 19.9%. If one adds to this the 200,000 or more who have emigrated since 1981 then one has a more accurate measure of unemployment; almost one in three. It is a clear statement that strategies for economic development have failed. They have failed because the preconditions for indigenous development have not been established. These preconditions are more sociological than technological because innovation is as much a sociological phenomenon as a technological one. In this paper, I wish to explore these preconditions albeit in a broadbrush way and the factors which have led to their absence in the Irish context. Odd though it may seem, in looking at economic development and at strategies for economic development in Ireland, a very good starting point is somewhere in the twelfth century B.C. Even from as early a period as this, one can see characteristics of the classic pattern of regional economic development and incidentally the irrelevance of the notion of peripherality.

In neolithic times, Irish society had reached a high level of sophistication in its technology and artistry, including its astronomy as can be seen in the magnificent tombs of the Boyne Valley. These date from about 2,500 B.C. There then seems to have been something of a decline and Ireland became what is commonly meant by a peripheral region - it was an exporter mainly to Northern Europe certainly of metals and possibly of other primary but more perishable products.

The archeological record shows that from about the twelfth century B.C. skills in metalwork began to develop. There is evidence for example of a travelling journeyman and of a metal-forming operation in a small settlement. These skills were undoubtedly acquired from other countries by the immigration of skilled craftsmen. New types of products, utensils, weapons and personal ornaments in metal mainly bronze and no doubt also in more perishable materials were brought into the country. These were copied, the skills to do so now existed, and were gradually improved until we see in the eighth century B.C., the late Bronze Age, an industrial revolution at its peak of prosperity. Irish products became much sought after across Europe for their technical and artistic excellence. The imported technology had been improved through innovation and upgrading of the domestic skills resulting in products more advanced than elsewhere. The international trade can be seen in for example the skull of the Barbary Ape found in Navan Fort (Emhain Macha) in Co. Armagh or the cargo of Irish products found in a shipwreck at Huelva in Spain. It was a Golden Age in which Ireland prospered even though it lay to the far north-west and very peripheral to the Eastern Mediterranean which most would consider the core region of the time.

Decline followed, the Celts invaded the island with the Irish finally dominating and with their dominance the country developed a new confidence and again became open to external ideas, cultures and technologies. From the fourth Century A.D., the archeological and historical record shows importation not only of Christianity but also of new metalworking and other technologies and of new artistic and decorative forms. These can be traced in origin to many different places such as Rome, the Elbe region of Germany and Syria.

These new cultures, technologies and skills were absorbed into a unitary indigenous culture which had confidence in itself. It was a society open to inflows of information from outside and willing to acquire new skills and design techniques and to absorb these and a new religion confidently into its own culture. The internal networks were such that there was rapid dissemination within the country, and these were intensified as the monasteries and universities developed.

Technological and artistic innovation and creativity resulted and Ireland reached a new level of self-generated economic wealth and prosperity. In this second golden age, one could almost describe Ireland as the first information economy. It became not only the centre of learning of Europe with several large universities but the storehouse of much of Europe's literary wealth.

To cope with this role, there were innovations in manuscript reproduction and decoration; the first form of long hand with upper and lower case was invented. The metal working and design technologies reached new levels of sophistication. It exported its' learning, technologies and innovations east to Kiev and south to Italy. All Europe was taught to write in upper and lower case and the Germans how to make beer. Strabo, writing from Reichenau on Lake Constance, said that the Irish so seemed to have taken up wandering across Europe that there could be but few left in their island home. In fact of course there were many left and these were joined by thousands of students and no doubt craftsmen and others from the European mainland, intensifying yet further the information flows. For two centuries Ireland had an incredible influence, but it was a period without a dominating personality. There were many personalities. It was not the push of an empire but of entrepreneurs trading in knowledge.

In these two golden ages Ireland demonstrates the classic mode of transformation of a backward or primary product supply region into a self-generating economy. Japan, or rather the region of Tokyo where it all began, went through the same stages:

1. Acquisition of skills, technology and culture from outside sources
2. Their absorption into the local culture and enrichment of it
3. Replacement of imports by locally made products
4. Improvement of the technology and design
5. Innovation
6. Exports of products
7. Exports of technology

Regions and localities which generate their own economic prosperity through new economic activities based on innovation have the following characteristics:

1. A distinctive culture, a technical culture which combines the technological and the aesthetic in a unitary culture, is confident in itself and therefore an open culture
2. Intensive open networks internally within the region through which information and ideas in technology, design, art, literature, religion and philosophy are easily passed; very low internal barriers
3. External linkages and networks through which in the first instance information and ideas are obtained usually by first importing the products and then the skills and other information and ideas

These networks then become two-way for trade and ideas as the region becomes a prosperous self-generating economy and enters on a process of continuous reinforcement of its creativity and innovativeness, by importing the innovations of others, improving these and creating new ones and then exporting these further innovative products. And so the process goes on, each self-generating region building further on the technology, design and ideas of others.

It must be emphasised that the technical culture must exist or be put in place first so that the internal networks intensify and develop an open confident character. These internal networks must exist so that the external channels lead to a flow inwards rather than draining outwards.

The three characteristics are the necessary preconditions for innovation and self-generated economic prosperity.

Economic prosperity is generated and maintained by the creation of new economic activities. They create new wealth and employment because they are innovative. They manufacture a new product using a new technology, they find a new use for an existing technology and introduce a new way of meeting an existing or latent market need. They replace declining economic activities and absorb idle resources. The new wealth which they create has a cumulative effect through further purchasing and investment.

New economic activities most often take the form of new firms founded by independent entrepreneurs, whether as individuals or in small teams. Once the process of new firm foundation is started, it gradually becomes cumulative and self-reinforcing. The foundation of one new firm leads to another. One firm spins off another and so on. A few grow large and create major new employment and wealth and yet further opportunities for new firms.

The start-up of new firms is the central pivotal activity of economic growth and renewal and the act of start-up is the real innovation. It is through this act that innovative products are brought to the market place. Innovation and the creation of new economic wealth are entrepreneur-led and are dependent on the entrepreneurial vitality of the economy, on having many small personalities, and on exchanging information and ideas with one another.

Entrepreneurial vitality and therefore the creation of new economic wealth are local. They are less than national and even less than regional. They happen not in general but are brought about by specific people in specific places. For these people to bring about these new firms, economic activities and innovations, the three characteristics of technical culture, intensive internal information networks and external linkages must be in place.

Where these do not exist, die or are depressed, the local economy will not generate entrepreneurial vitality. In an economy of low vitality, when developing programmes and strategies for technological and industrial development one risks another phenomenon. The process of entrepreneurial vitality once started is dynamic. It becomes a cumulatively self–reinforcing upward spiral. The process of decline is also dynamic and cumulatively self-reinforcing. This means that in a depressed economy unless development strategies focus first on the preconditions they will tend to reinforce the weakness and therefore the process of decline. This point cannot be over-emphasised.

Peripherality as such is not a factor. Ireland in its first two golden ages was even more peripheral geographically to what most would consider the core regions of the eastern and central Mediterranean than it is now to what is called the core of Europe. It is no more peripheral than Japan or Finland, and there are more central places such as Liège and Hamburg which are in decline. The key factor in the concept of peripherality is not geographical location but control from another region. This Ireland has had to an exaggerated degree. Since independence it has failed to develop the essential preconditions and, if anything, has seen an increase in the level of control from other regions.

The story is a long one. The second golden age in Ireland declined in face of the raids by the Vikings and then the struggles with the Viking kingdoms established on the coast. These were gradually absorbed into Irish culture and refreshed it by the further intake of information and ideas. Irishmen began again their wanderings across Europe and Ireland entered into a further period of growth in design, technology and products. This was however brought to a sharp halt by the invasion of the Anglo-Normans in 1170 and the destabilisation which followed. Eventually in 1603 Ireland was finally conquered although two further major wars were to take place in the 17th century.

From the point of view of this paper the pattern of colonialisation which followed had three features of importance:

1. The language, laws, customs, literature and religion of the Irish became illegal.
2. No Irishman could become a skilled craftsman.
3. Ownership of property and improvement of property were virtually impossible because improvement was heavily penalised by increase in already punitive rents.

The effect of these measures was that the technical culture of the country was destroyed, and its indigenous restoration through adoption of new techniques or technology was prohibitive in its consequences. Irishmen and women still crossed to Europe but the information and ideas which returned tended to the political. There were no internal or external networks for the technical or technological except through the English settlers or Anglo-Irish as they became known. The innate culture of the Irish, however emaciated, did survive and gradually an educational system emerged, some of which was visible in private schools but more was in what were called the hedge schools. At the end of the eighteenth century Ireland had the remarkable achievement of a higher number of its children at school than any other country in Europe.

The orientation of the education was however to the classics and humanities. Any incipient trend to develop industry in the earlier part of the nineteenth century did not survive tax, shipping and other measures imposed to protect English industries from possible competition. Development of craft (professional) skills was again restricted and the orientation to the humanities was reinforced when the English government and the Irish Catholic Church agreed in the 1860s the shape of the national educational system. The orientation of the Irish and naturally enough, of their classically educated clergy was to the humanities, the classics and the academic sciences. It matched the class based orientation of the English educational system with its separation of a humanities and academic culture from an inferior technology and "trade" culture. The university system, the Queen's Colleges established in the later nineteenth century, was modelled on the English and rooted in this two-culture syndrome.

A consequence has been that Ireland has absorbed many of the features of an economy in decline and an industrial power relying increasingly on price for its competitiveness.

The Irish educational system has not changed fundamentally since these events. It retains its weak technological orientation. Nevertheless the general level of education is undoubtedly very good and there is a relatively high participation rate in secondary education as can be seen from the numbers taking the Leaving Certificate (Table 1).

Along with other measures to improve the technical infrastructure, eight regional technical colleges were established in the late 1960s but some show a distinct tendency to aspire to academic qualifications rather than the kind of thorough professional skill training which distinguishes the apprenticeship/vocational training systems of more prosperous countries. Two university level business and technology institutions known as National Institutes of Higher Education have been set up in Dublin and Limerick.

There has been a relatively heavy investment in micro-electronics, electronic engineering and computer science/software in universities and more recently biotechnology centres have been established in the universities. Undoubtedly, the quality of many graduates is high as is clear from the number working in research for multinational high technology firms, as for example in the Eindhoven Laboratories of Philips. Nevertheless, the technological orientation of the educational system remains weak. Engineers are aca-

Table 1: Average percentage of children taking examination and entering third level education in any one recent year.

8.0%	Left without taking any examination
8.0%	Failed group or intermediate certificates (age 15-16)
28.0%	Passed the group certificate
57.0%	Passed the intermediate certificate (some passed both)
58.5%	Continued to and took the leaving certificate (age 17-18)
25.0%	Entered third level education
Of whom	
39.0%	Entered universities
51.0%	Entered RTCs and NIHEs
9.0%	Entered teacher training colleges

demic not production–oriented. As one professor of engineering has said "our job is to produce research engineers". The apprenticeship/vocational training system was never strong. In 1983, there were 3,633 first year apprentices of whom almost 2,000 were in construction and motor mechanics and 694 in the electrical trade.

Inevitably, the ambition of parents is to have their children enter university. Education per se is the perceived objective. It is not perceived that it is an education in a half-culture. That understanding of technique and technology which leads to the creation and appreciation of works of art and to the design and manufacture of functionally useful, quality and visually beautiful products is the missing half of a unitary culture which once existed in Ireland. It still exists in the self-generating regions of Europe, those places which maintain and renew their prosperity. The creativity of the Irish finds expression virtually only in the written and spoken arts in which it has a standing beyond the numbers of its people.

Ireland's focus on electronics, computer science and other "science" based industry is shared with England and indeed the United States. The Financial Times has on more than one occasion remarked on this syndrome that it is "respectable" to start up or work in a new technology enterprise. These have grown in England while traditional industry which is not "respectable" has declined, as it has in Massachusetts. This narrow focus on "science" based technology and industry has developed interactively with the Industrial

Development Authority's policy of attracting foreign firms in such industries to invest in branch plants in Ireland.

Academic in its origins, this narrow focus flies in the face of the realities of self-generated and maintained economic prosperity. "High tech" industries in the U.S. account for about 5% of employment much of which is low-skilled. High-tech very rapidly becomes low tech assembly especially in the American corporate culture. The U.S. model is of wealth and employment creation in a new sector built around a major innovation. Such major innovations and new sectors tend to be introduced at the peak, more or less, of a long wave of prosperity, but the next wave of prosperity is pushed by the diffusion of this innovation across many other sectors creating a wave of innovations. For this trans-sectoral innovation to happen is required entrepreneurial vitality in a diversity of sectors and the primary precondition for entrepreneurial vitality is a strong technical culture. In these modern times, this is dependent not on the university education system but first on the creation at the primary and secondary levels of the education system of a wide ranging aesthetic and technical appreciation of design, quality and how things are made and used. Included in this understanding must now be computer literacy and the role of biotechnology. Secondly, success is significantly dependent on the apprenticeship or professional vocational education system. In the regions of Europe there is no correlation between the economic prosperity of a region and the numbers of students as a percentage of the population at university. There is a strong correlation between the numbers of young people in apprenticeship/professional training and the quality of this training. In Switzerland for example 50% of young people go through some form of apprenticeship system and of the other 42% who emerge into adulthood with a qualification many are also technologically rather than academically oriented.

In regions of Germany, Northern Italy and Denmark one can observe the same phenomenon. The training is not narrow. It gives each participant the skills and ability to understand the entire process before acquiring a specialisation in part of it. It is thus possible to add on at a later stage further skills as for example in programming. The ability to understand the entire process enables one to see the significance of a new technique or customer need, in other words to innovate. The skill training is wedded into the traditional culture of these prosperous regions each reinforcing the other. They have not lost that "aesthetic faculty, once the possession of all countrymen" which according to the English writer H.J. Massingham (1945) was what suffered most from the impact of mass production in England! It is one aspect of the decline in English technical culture, "the skills of the craftsman of Sheffield" which Adam Smith extolled as the core of England's industrial supremacy. The numbers of apprentices in England have strongly declined and the quality of their training inevitably weakened. The causes are sociological not technological.

A German manager of an engineering company once remarked of the German apprenticeship training system, that it did not matter if a young person who emerged as a master baker did not bake a loaf of bread again. He or she had acquired a sense of quality, an appreciation of the well made article, and an attitude of servicing the customers needs. Whereas a television critic of the *Financial Times* amusingly saw the loss of the aesthetic faculty in the participants of the forty or so quiz shows featured each each week on British Television, in that one of his four or five conclusions was that they were dressed in clothes no Italian working class people would be seen dead in.

Aesthetic appreciation is a very necessary component of technical culture. Design is as important as technology in innovation. It begins with visual appreciation. No one would doubt the visual quality of many Finnish, Danish and North Italian products. They come from a unitary technical culture in which the technological and aesthetic are mutually supporting and inter-acting. Newer technologies, higher skills, have been absorbed into the traditional culture and have enriched and enhanced it.

In Ireland it is widely agreed that the aesthetic visual quality is lacking. This lack is primarily a result of a weak apprenticeship vocational training system, which inculates the appreciation of the well made article, well suited to its end purpose. That it can be developed is amply shown by the development of Waterford Glass. The original Waterford Glass company was squeezed out of existence by the oppressive tax and trade regulations in the 1830s. It was revived in the immediate post-war years by bringing in skilled Czech glass workers. Training and a strong apprenticeship system resulted in the creation of an international name for quality and a company successfully growing and competing in what was and is a highly competitive market in which design and quality, the skill intensiveness of the product, determine purchasing decisions.

The significance of technical culture in entrepreneurial vitality and the creation of economic wealth is that entrepreneurs found their firms around their own skills. An educational and training system which produces a large number of young people who have technical skills to a high standard is producing people who can create their own jobs. They can with further experience and more and diverse information and know-how become entrepreneurs. Only a minority do so, but equally importantly they have available a workforce which can produce skill-intensive products to customer satisfaction and can themselves innovate. They have a total culture. Where there is a large number of skilled people in a diversity of industrial sectors but with a strong engineering industry and relatively large number of small firms, one finds that the local economy is self-generating and self-renewing. More new firms are being founded and there is a higher level of innovation. This is a result of the intensity of the local information flows, their openness and the willingness to give as well as take. It is possible within half an hour to meet someone face to face to get information which one needs or an introduction to the one who has it or to obtain a decision. There is a responsiveness to need. The entrepreneur or potential entrepreneur is kept continuously up to date on what is happening in his own sector though frequent personal contacts. He is also familiar with what is happening in other sectors. The skills of one sector are applied to the problems of another. Most innovation occurs from trans-sectoral flows and diffusion. One company and one sector localities eventually go into decline. The information flows lack the diversity to support innovation and entrepreneurial vitality and therefore the creation of new economic activites (Sweeney 1987).

These internal, open and intensive information flows are vital. Various studies in the U.S. have focused on why large American cities with similar resources in universities, numbers of engineers and scientists, expenditure on research, banks and other financial institutions differ so markedly in their entrepreneurial vitality and therefore in their ability to renew their own prosperity. There are few new firms being founded in some. Others have a high rate of new manufacturing business formation.

The differences are due to the weak internal flows of information in the cities without vitality and the openness and willingness to give information and other resources in the cities with vitality. Wherever one looks, one sees that high vitality is associated with frequent and intensive information contacts, whether in the Prato Textile Basin in Tuscany, the small firms of West Jutland or the new technology firms of Cambridge, England.

Much has been said about the Cambridge Phenomenon. It provides a sharp silhouette of the preconditions for entrepreneurial vitality albeit in a narrow sector. Cambridge University has a strong technological orientation with over 50% of its students in science and engineering. While much of the scientific research is fundamental it is still oriented to eventual transfer into technology. The staff have strong links into industry and business and are active in policy, business and industrial affairs outside the university. There is an undoubted confidence in their own culture and its breadth. New firms started to be founded in the early 1960s because the potential entrepreneurs found encouragement, responsiveness and support in the provision of resources. More new firms were founded, spin-offs occurred, key staff moved from one firm to another. The networks of contacts between the families of firms which grew up around Cambridge were intricate and intense and spread out to banks and other resource-providing agencies in the area. When one looks at the few other universities which have seen clusters of new firms grow up around them such as MIT, one sees there also the open and responsive networks founded in a confident technical culture. One cannot as yet see these developing in universities in Ireland, but in this they are similar to most universities elsewhere. It is also worth noting that the Cambridge Science Park was established after the preconditions were in place and the process of entrepreneurial vitality was already dynamic. Growth of the entrepreneurial firms and sector has however become frustrated more recently due probably to lack of diversity, the low technical orientation generally in the wider business environment and increasingly external investment in the Cambridge area and firms.

The intensive networks are significant in another context. The previous experience of most entrepreneurs is in smaller firms. In smaller firms one is not only in more day to day contact with the other functions in the firm and therefore able to see the total business, one is in contact with suppliers and customers and therefore aware of technological developments and market trends.

Identification of an innovative opportunity is much more readily made and becomes more probable as more new firms are founded, clusters grow larger and overlap clusters in other sectors.

In Ireland, these internal dynamics, the intensive networks between small firms, between small firms and large firms, between potential entrepreneurs and between potential entrepreneurs and customers have not developed. The technologically weak educational and training system in not producing large numbers of young people who have the skills to create their own jobs is the primary cause. The failure can be measured by contrasting Ireland with regions which have generated their own prosperity. The barriers can also be measured in the structure of employment in Ireland (Table 2), as it was in 1961 and has developed since.

The shift from a technically weak agricultural sector has not been to a technically strong manufacturing sector, it has been to the technically weak services sector. Indigenous manufacturing industry has also declined. According to a recent study (Hart 1989) the picture of employment loss and growth is as follows (Table 3):

It is now well established that branch plants interact weakly with the local economy and depress the capability for self-generated growth. Ireland has been no exception. Indige-

Table 2: Employment: Ireland '000s

	1961	1971	1981	1987
Agriculture and mining	388.2	283.5	199.7	164.0
Manufacturing industries	179.4	213.6	238.1	228.9
Utilities	10.2	14.2	14.4	
Building	59.6	84.5	102.2	71.3
Services	415.0	459.0	583.4	623.5
Total employed	1,052.5	1,054.8	1,137.8	1,087.6
Unemployed	42.5	49.5	113.0	242.8 (1988)

SOURCE: Census of Population

nous firms have existed in increasingly weak information networks and in an environment which does not provide a potential for innovation. The problem can be seen more starkly in two subregions, the West and the Mid-West (Table 4). The Mid-West has been the location of intensive and imaginative programmes by the only subregional development authority in Ireland, Shannon Development.

In 1982, 64.8% of manufacturing employment nationally was in indigenous firms, 47.7% in the Mid-West and 50.3% in the West. Indigenous industry is thinly dispersed and declining. Manufacturing employment has not increased to any significant extent over the 26 years since 1961, although the West subregion has improved due almost entirely to foreign branch plant employment. The environment is necessarily weak in its internal flows and technical culture.

Table 3: Manufacturing Employment Gains and Losses 1973 - 1986

Companies established by 1973 lost employment by 1986	
• Indigenous firms	64,300
• Foreign firms	28,700
Companies established since 1973 gained employment by 1986	
• Indigenous firms	30,000
• New Foreign firms	43,100

Table 4: Employment in two sub-regions

		Agriculture	Utilities	Manufacturing industries	Building	Services	Total
Mid-West	1961	44,959	826	24,038	6,635	37,371	113,829
	1981	32,622	1,187	25,430	12,969	53,610	123,823
	1987	20,200	22,800		6,600	47,700	97,300
West	1961	66,882	408	5,854	4,416	25,756	103,316
	1981	30,628	724	13,980	7,286	37,906	90,524
	1987	26,200	14,900		5,900	46,000	93,100

Existing in this environment means also that there is not a demand/pull for quality. Hitchens and O'Farrell (1988) compared the products of small indigenous firms in the Mid-West and North of Ireland as well as regions in Britain by asking matched companies for their opinion on quality. The Irish products were felt to be of poor quality by firms in Britain, especially in South-East England, while firms in Ireland tended to find all products satisfactory. The age of equipment in Irish firms tended to be newer than those in Britain, indicative of the grant basis of industrial development.

Casey (1988) in reporting a study of the uses of different sources of information by foreign and indigenous firms demonstrated the weak links of indigenous companies (Table 5). They used suppliers of equipment rather than original manufacturers, a long observed phenomenon in less-developed regions. They seem to make little use of one another and have a heavy reliance on the two state services.

The barriers to development of intensive information networks among indigenous firms in fact seem to be increasing. Although over 3,000 firms founded since 1973 have survived, their average employment is only 8.1. Growth is difficult.

These barriers are being enhanced by the way in which external linkages have developed. These tend to drain the country rather than feeding ideas and information into it. Control in effect still lies outside the country. This is a deep rooted sociological phenomenon and to understand it one must look at another major factor in the creation of modern Ireland - the massive emigration which began in the great famine of the late 1840s and continued until the 1960s. The Census of Population of 1971 was the first to

Table 5: Percentages of indigenous and foreign firms using different cources of technical advice/assistance

	Indigenous Firms	Foreign Firms
Equipment suppliers	42	53
Equipment manufacturers	30	56
Parent/sister company	4	75
Other company	7	9
University	6	15
Regional technical colleges	6	11
National institute of higher education	4	14
College of technology	3	7
Consultants	10	12
EOLAS (Science and Technology Agency)	46	48
FAS (Training and Employment Authority)	32	42

record a net increase in population since that of 1841. Resolution of economic difficulty for a population without skills lay in emigration even although what awaited very many of them in the nineteenth and early twentieth centuries was harsh.

Nevertheless, this migration created networks and linkages extending across the world. Through them opportunity was perceived to be greater elsewhere. Most of the firms founded by Irish entrepreneurs have been founded overseas in sectors easy of entry such as the construction industry.

Inevitably, internal networks were weakened with the actual flow of people and the looking outwards. After independence in 1922, tariff and nontariff barriers were established to protect Irish industry with the objective of nurturing its growth. There is nothing wrong with this approach in principle if programmes to develop the technical culture in all aspects are put in train simultaneously. This was not done, and the tariff and other barriers served principally to block the inflow of technology and other information and ideas. The four year economic war waged by Britain until 1938 caused considerable economic damage as well as wreaking hardship. It was followed by the Second World War. Irish industry failed to recover and there was a tragically large surge of emigration in the late 1940s and 1950s, renewing and enlarging the contacts overseas.

In 1958, the first plan for economic development was published. There followed wide ranging programmes to develop the technological infrastructure and industry. Confidence began to develop. Gradually the net migration rate became positive and as a result of people returning home and staying home the population grew strongly to give it a young profile (Table 6). The number of young people at school after the age of 15 grew from 78,999 in 1961 to nearly 200,000 in 1981.

Table 6: Population by age groups

Age	1961	1981	1986
0 - 14	877.3	1,037.9	
15 - 24	391.9	608.4	
25 - 44	635.2	839.2	
45 - 64	598.9	589.9	
65 +	315.1	368.0	
Total	2,818.3	3,443.4	3,540.6

Confidence in Irish culture grew strongly. Music and literature blossomed. An Irish publishing industry and not merely writing industry came into real existence. But in matters technical and industrial, the culture was outward looking and strongly dependent. Irish products were regarded almost automatically as of poor quality. There was little support of Irish maufacturers who attempted to improve their range. The problem was bound up with the colonial past. The vast majority of Irish exports went to Britain and Britain was the major supplier.

The dominant influence of Britain was reinforced not only by the large number of emigrants but also by an administrative system which tended to follow developments there. Inevitably, like Britain, Ireland was a market dominated by price, and quality was a trade off against price. There was no demand/pull, or supplier push, for quality. There was no feed into Ireland of information which would develop the technical culture. Linkages to more quality demanding markets in Europe were not developed, and this was at least in some degree due to the poor quality of language training in the general education and university system.

Irish manufacturers therefore tended to compete in the lower end of the market, the most vulnerable to competition both from a low wage Britain and from NICs. Inevitably, since free trade was introduced first with Britain in 1965 and then the EEC in 1973 there has been a decimation of larger firms in traditional industries such as textiles, clothing, footwear and furniture.

In spite of the expenditure on creating a technical infrastructure, the main thrust of development policy to support indigenous industry was in the form of grants. Cash was and remains the main measure. Non-cash measures to make a direct impact on raising technology, skills, and quality have been minimal.

Industrial policy was heavily oriented to attraction of investment by foreign companies, mainly U.S., and in this, there can be no doubt, the Industrial Development Authority is a highly professional organisation. There are now some 800 foreign owned firms employing about 86,000 people. There was a priority need to create employment in order to turn the tide of outward migration. The mistake was in the heavy dependence on this strategy.

In it, the IDA inevitably strengthened the outward linkages of the Irish economy but in a way which depressed the internal networking. Branch plants of themselves transfer local inputs to the external parents. As can be seen from Table 6 they have stronger links than indigneous firms to universities and other third-level institutions, again reinforcing the flow of information outwards rather than inwards to the internal networks. Even when there was an opportunity to assist an Irish firm to undertake subcontracting, a foreign firm was sought. The attitude of dependency on the external world engendered by migration, colonialisation and the failure to create self-generated prosperity was endemic.

Conscious of the need for improved design of Irish products the Export Board established a design workshop, bringing in Scandinavian designers. In medical parlance it was invasive. There have been some good products but inevitably it has had little influence in recreating the aesthetic faculty, the appreciation of the well made and functionally designed article. Design is still largely treated, as it is in Britain, as something separate from technology and manufacturing. The technologies of materials and shaping materials are not an integral part of the teaching of design nor vice-versa.

The state system is large and dominates instead of supporting from below. It tends to create and maintain linkages to itself rather than intensifying the critical networks between indigenous or quasi-indigenous firms. This dominance is reinforced by a heavily centralised administration which again weakens the local linkages. The linkages tend to be to Dublin, and have been accompanied by a marked shift of population to the capital city, a feature characteristic of centralised economies. Decentralised and federal economies perform economically much better than centralised economies because they do allow the local networks to become dynamic.The dominance of the state system has led to a dependency attitude that government will create jobs. It is an attitude which inevitably leads to bankruptcy.

The state technological and research institutes established from 1958 on eventually seemed to lose their way, and have recently suffered massive cutbacks in staff. That they missed their way was largely due to the heavy reliance of the state on an inward industrial investment policy. The other side of this particular coin was the absence of a technological policy and programmes to develop indigenous entrepreneurship and industry as the other and long term industrial development policy. It would have required a firm recognition that technology is needed in all sectors of industry and that providing the skills to create one's own job is the sine qua non of entrepreneurial creation of new economic activities and firms. In the absence of such a policy a number of events happened which in effect reinforced the decline of the technological institutes.

This brings me back to an earlier point. Decline, like growth, is a dynamic cumulatively reinforcing process. Specific actions adopted in a locality or region which has the necessary proconditions further reinforce the process of growth. In regions without the preconditions, they may well reinforce the decline. Hence a programme which reinforces the external linkages of academics who are weakly linked into the indigenous networks will reinforce the outward flow of information and retard local innovation. This phenomenon has been well studied in the past (Crane 1974). The Esprit and other programmes in which Irish academics have dominated in Irish participation will in practice have a negative impact on the capability for self-generated development in Ireland. A similar comment has been made about the building of "powerful broadband ISDN communications highways into less developed regions" (Casey 1988). They might well be "a new form of emigration". The empirical data in fact shows that installing a telephone into a high percentage of homes in less developed regions is a much more effective way of stimulating indigenous development (ITU).

Such a programme would lack glamour however. One has to make such a comment in the context of programmes which are not geared to the specific conditions of less developed or declining regions whether they are peripheral or more centrally located. Ignoring the absence of the preconditions will lead only to further decline in the capability for

Table 7 Net migration 000's 1971 - 1987

1971 - 1981	103.9
1980	2
1981	-1
1982	-14
1983	-9
1984	-20
1985	-28
1986	-27
1987	-32
1981 - 1987	-131

self-generated prosperity. Persistence in using the notion of peripherality itself only postpones more effective policies.

The final statement on the effectiveness of strategies for economic development in Ireland lies in the rate of unemployment which in August 1988 was just under 20%. The official figures for migration are given in Table 7. The unofficial estimate of illegal migrants in the U.S. is 150,000 and the number of migrants in Britain, Australia and Europe since 1981 cannot be much lower.

Possibly the most remarkable phenomenon is that the private sector has remained virtually inactive while its market literally walks away.

Ireland lacks confidence in itself. Until it can recreate a technical culture through a highly skilled population it will not gain sufficient confidence in itself to generate its own prosperity. The cycles of unemployment and emigration will be a recurring phenomenon until Strabo is right, there will be none left in the island.

References

Casey, T. (1988)
Technical information support to manufacturing firms within the less developed regions of Europe. In: Information resources and corporate growth. Punset, E. and Sweeney, G.P. (eds.), London, Pinter.

Crane, D. (1974)
An interorganisational approach to the development of indigenous technological capabilities. Paris, OECD.

Hart, M. (1989)
Entrepreneurship in Ireland: a comparative study of Northern Ireland and the Republic of Ireland. Journal of Entrepreneurship 1(1).

Hitchens, D.M. W.H. and O'Farrell, P. (1988)
The comparative performance of small manufacturing companies in Ireland. Queen's University, Belfast. mimeo. ITU and OECD (1983). Telecommunications for development. Paris.

Massingham, H.J. (1945)
Introduction to: Thompson, F.: Lark rise to Candelford: Oxford Unversity Press, Oxford.

Sweeney, G.P. (1987)
The entrepreneurial firm as a learning system in the information economy. The Information Society, 5: 91-99.

5.3 Problems of an Industrial Region in Decline: The Case of Birmingham

Harry Nicholls

1 Birmingham in Decline

Regional problems are very deep seated and bound up with the history of a region. Therefore, to understand what is presently happening in Birmingham, we have to look at its recent history. The City of Birmingham is a city of about 1 million people, situated within a region of about 7 million people within a one hour drive from the city. For almost 200 years Birmingham was extremely prosperous. But in a 10-year period from 1970 to 1980 Birmingham moved to having 21% unemployment, 34% unemployment in the core area and if the male unemployment for the age group of 16 to 24 in the inner city area is looked at, it was something like 70% or even 80%. Yet only 10 years earlier, in the 1960s, unemployment was less than 1%. Within that 10-year period, 190,000 jobs were lost, more than for the whole of Wales and Scotland put together, equal to the manufacturing population of the State of Western Australia. Within seven years, manufacturing employment in Birmingham, and manufacturing output dropped by 34%. Capital investment in manufacturing dropped by 36% in the same period, 50% faster than in the U.K. as a whole. By 1981, the census statistics showed that Birmingham had the largest and most severe concentration of deprivation in the United Kingdom, indexed at 49%, 20 times worse than the expected figure.

2 Reasons for Decline

What were the reasons for this sudden decline? A major cause was the failure of the City of Birmingham and the West Midlands as a whole to structually adjust its economy. The economy of Birmingham dates back about 200 years, to people like Matthew Boulton, James Watt and William Murdoch. Birmingham at that time was the centre of high technology in the world. But high technology in those days was not microchips or bio-technology, it was steam engines and gas lighting. Later, Birmingham built its economic strength on metal manufacturing of various kinds and then on vehicle production. Birmingham was the centre of the British motor industry. But, for a number of reasons Birmingham fell into decline.

2.1 A Loss of Entrepreneurship

One of its strengths was its concentration on particular industries which led to the development of all the infra-structure services which supported them. It also depended very much on a particular approach to entrepreneurship and wealth creation. But the fundamental truths, which go back to Boulton, Watt and Murdoch, namely to apply the best thinking in science and technology to manufacturing, were forgotten, what is described by British people quite often as that mundane process of wealth creation. What was forgotten is that it is that mundane process of wealth creation which allows for all the qualities of civilised life. Over the last 30 or 40 years, when Birmingham began to start losing its jobs, it had moved away from the idea of applying the latest thinking to wealth creation. Simple pragmatic approaches to business prevailed, which neglected investments, research, and the searching out of new markets, resting on its strength in selling in comfortable, protected markets.

2.2 Relocation of Industries

Regional policies played a particular role. When Birmingham had less than 1% unemployment in the 1960s, central government refused to allow industry to expand, and adapt and grow naturally within the West Midlands. It was forced out to depressed regions. One example is the Rootes Motor Company in Coventry, Birmingham's neighbouring city, which was a major car producer in the 1960s and forced out to Linwood in Scotland. But the subtle infrastructure for the supply of component parts and all the rest of it, and the particular attitude of engineering skills which are required in mass manufacturing, were not present in Linwood. As a result, twenty years later the Linwood plant is devastated, Rootes disappeared into Chrysler, Chrysler disappeared into Peugeot. The operations were moved back to its traditional home in Coventry, the Peugeot plant in Coventry being one of the most productive Peugeot car plants in Europe with high investment levels. The industry is growing again, building on established strengths, located in a sensible place in relation to the manufacturing skills of the workforce and also other economic factors.

2.3 Changes in Attitude and Industry

Some of Birmingham's problems were self-inflicted because the City of Birmingham itself, once it got rich, did not like all this dirty industry around, and was encouraging dirty industry and small scale industry to move out. Birmingham is a paradoxical city, it is known as the city of a thousand trades and as a city with masses of small firms. But 40% of its employment is accounted for by 10 firms. There are 10 big companies and then lots of little ones. But when the little ones had grown up, they had grown up just naturally intermingled with housing and so on. The city planners attitude was to have industry in one place, houses in an other, shopping here, entertainment there. Small firms scattered around the city were an obstacle to this, so they were encouraged to move out, which meant that costs increased.

Also new problems arose. Birmingham's industry was devastated. The car industry lost tens of thousands of employees in Birmingham and component companies lost tens of thousands, partly because of failures of marketing but also because of changes in technologies. Automation means that fewer people are needed. Change within manufacturing technology and manufacturing systems, for example, "just in time", means that component manufacturers have got to move closer to where the assemblers are. So, as Birmingham assemblers of motor cars were weakened, partly by their own myopia, partly by government policies, also the component manufacturers that depend on them have had to rationalize to survive, have had to re-address and, in some cases, have had to move to another continent. So today a Birmingham firm, GKN, producing the propshafts of frontwheel drive motor cars are the lead tenants on the Oakland Technology Park just outside Detroit.

2.4 Centralization of the Economy

In the last 30 to 40 years, the United Kingdom is becoming increasingly a centralised economy. With the communications revolution there has been a transportation revolution and, given the highly efficient Stock Market in London, resources are allocated very effectively, but in a very neutral fashion. As a consequence, local investment has disappeared and with it local stock markets. Financial investments in the U.K. are all channelled primarily to London, some to Edinburgh, and then are invested where the best rates of return are achieved, which can be anywhere in the world. So there are no local loyalties with money.

Furthermore, the U.K. is suffering the problems of being a "brand economy". Those companies which were sufficiently successful, have become national companies or multinational companies and have tended to move their headquarters out of the region. The U.K. is a small country. Birmingham is only 90 minutes from London by rail or two hours away by road. It is very easy for companies to move their headquarters down to London. With them they move their loyalties to the local labour force. The kind of solutions that worked in places like Baltimore or in Boston in the United States are probably not possible in places like Birmingham because the local industrial leadership is missing. And it is missing because it has moved to London, leaving the branch managers without the power of investment decisions. So companies no longer have the identity with the region. As a consequence in the present governments's enterprise economy, which after all was the fastest growing economy in Europe apart from Spain in the 1980s, sometimes the enterprise has to come from local authorities. Unfortunately, with the British central government at the moment, the only kind of enterprise which is not encouraged is local authority enterprise - because that tends to be Socialist rather than Conservative.

3 The strategy of the City of Birmingham

Nevertheless, the City of Birmingham took the initiative with a six-point strategy consisting of:

1. A business development programme, putting in advice, concentrating on support systems for small-scale local companies. The large ones and the multinationals can look after themselves.
2. Money is being spent on promotion, because it was realized that Birmingham has an image problem. People do not think of Birmingham as being an attractive city in which to live and to work - they see it as a dirty, old, industrial city - they do not realise that with the decline of industry all of the dirt, smoke and congestion has gone as well. That has involved, for example, such promotional activities as bidding for the Olympic Games. Birmingham never wanted to win the Olympic Games, it only wanted the publicity that came from competing for them.
3. Action has been taken on technology. The City authorities were instrumental in forming Aston Science Park in partnership with Aston University and Lloyds Bank. Public/private sector partnerships were seen as very important for regional economic development, for developing technology through technology transfer centres, building on the power of the technological insti-

tutions and of the universities and spreading best known practice around the region.

4. An important task was to reclaim some of the physical dereliction which 200 years of industrialisation had left behind. In 1981 approx. 40% of Birmingham's industrial premises were over 80 years old. The pragmatic attitudes of local businessmen which had produced success in the past, also meant that at one time in the 1980s, part of the re-equipment of small-scale Birmingham industries meant buying 35-year-old capital equipment from other companies, and using it to replace 55- or 60-year-old capital equipment. So, advice was needed on technology to apply major technology, but also buildings were lacking. There were acres and acres of dereliction, and not wanting to end up looking like Detroit or certain other North American cities, money needed to be spent on putting that dereliction right.
5. Promotion for new investments was needed to encourage people to re-invest, as well as new training and community initiatives. If industrial structures are to be changed, then all the systems of training people will need to be changed as well, because otherwise people are being trained for the old technology. Sometimes, it can be hard to change training institutions .
6. It has been acknowledged that, given technological change, there are never going to be new car firms employing 20,000 people, or component companies employing 20,000 people, nor will there be major steel works or rod mills employing 10,000 or 15,000 people. Whatever would be done with the industrial structure, there would never be sufficent employment for all the people living within the city, and, therefore, it was sensible then to diversify into previously unheard of fields, such as tourism. Birmingham is now a major tourist centre but, with a difference. People don't come to Birmingham for the sun, they don't come for the water, even though there are more miles of canal then in Venice, all industrial canals now being turned into urban walkways. People come to Birmingham for business tourism, because one of the first major initiatives which the City of Birmingham took 15 years ago was to invest the equivalent of 70 million ECUs into developing the National Exhibition Centre. This was initially a project that flew in the face of conventional wisdom in the U.K. which says anything nationally, anything important has to be in London to be successful. The National Exhibition Centre in Birmingham is now one of the most successful exhibition centres in Europe, intensively used and doubling in size by the year 2000. It

gives direct and indirect employment to between 4,000 and 8,000 people. That is not a lot in the context of 150,000 to 190,000, but what has to be accepted, when trying to solve the problems of older industrial regions, is that there is no way to find a quick solution to the closure of a plant with 10,000 people. It is not possible to "magic in" another plant to employ 10,000 people. So, what can be done is to work on masses of smaller initiatives, growing little companies from the beginning, which can not be done quickly. But it is a much more effective process than trying to attract internationally mobile industry to come to your region, because internationally mobile industry which is attracted by grants and subsidies, will go off for the next best grant in two or three years time, it is no long term solution. A longer term solution is to grow your own but this takes a long time.

4 Summing up

To sum up, successful regional development must keep in mind:

1. To apply high technology and best practises to manufacturing.
2. To take advantage of the geographical and the industrial benefits of the area as well as existing skills and companies.
3. To allow local companies and start-ups to develop.
4. To encourage the formation of new forms of investment, with genuine venture capital, that invest in start-up companies.
5. To work in partnership with people starting new enterprises by supplying the help and advice they need.
6. To actively promote public/private sector linkages in all sectors.

5.4 Innovation-Oriented Policy in Regions with High Growth Dynamics: Three Winners in the Process of Structural Change - A Comparison of Baden-Württemberg, Massachusetts and Emilia Romagna

Jürgen Gabriel

1 Introduction

The general conditions for economic growth in all industrial countries have changed considerably since the mid-1970s. The most important changes hasve been: a steep rise in the prices for raw materials, which in turn led to unpredictable fluctuations in prices for resources; considerable changes in currency exchange rates; the appearance of new international competitors for mass consumption and investment goods; and, finally, a change in consumption patterns. Coupled to this, rapid progress in the development of technology has taken place. Due to closer international integration there has been an assimilation of national business cycles, and all industrialized countries have thus been faced with similar economic problems, such as slowing growth dynamics, unemployment, decline of traditional industries and budget deficits.

As a reaction to this tendency, increased efforts in economic policy directives have been undertaken in all countries. The policies have aimed at improving the basic conditions for promoting private investments, and internationally coordinated economic, monetary, and trade policy measures have been implemented. Since the beginning of the 1980s economic indicators in all industrialized countries have indeed shown a recovery of prosperity. On first reflection one could suppose that the necessary structural adaptations have been commonly carried through. But behind the macroeconomic data there are deep-rooted structural and regional differences.

Within national borders some regions have been more successful in dealing with economic and social problems than others. Many formerly prosperous regions have fallen deeply into crisis, while others have experienced a new boom. In the Federal Republic of Germany the tendency towards a north-south divide has increased. In Italy the gap between the poor southern regions and the rich north Italian industrial regions has become greater. People and capital in the U.S.A. have fled from the so-called frost belt to the so called sun belt.

Some economic regions have done extremely well and are considered to be role models for their particular countries as well as in international comparison. In the Federal Republic of Germany we are speaking here of Baden-Württemberg, in Italy the Emilia Romagna and in the United States Massachusetts. Therefore, many politicians and political advisors have called for research into the largely undocumented reasons for these successes so that the model can be transfered to less successful regions. To understand the similarities and differences in the process of success, a comparison of these successful regions has also been called for.

This paper tries to meet this demand in a first approximation. Theoretically and empirically strictly ascertained results and quick answers concerning a transferability of the "growth models", however, are not to be expected. First, the structural conditions and the conditions of implementation for policies are very different. Second, the method of comparison is necessarily too "soft". Strict causal relations are nearly impossible to isolate because of the various factors of influence. Nevertheless, the comparison made here offers explanations and hypotheses that can stimulate further research and future practice in the field of innovation-oriented regional policy.

2 Unfavourable Initial Conditions

The economic success of these three regions is indeed impressive against the background of prior prognoses. Before they set off to leave other regions behind, these were considered to be problem regions on the grounds of their alleged disadvantages. Massachusetts, for example, had an obsolete industrial structure, where shipyards, machine construction and the textile industry dominated, all being industries that had come under severe pressure from national and international competition.

The recession of 1973-75 hit Massachusetts especially in the form of rising oil prices, decreasing public spending, and unemployment which was above average. A creeping "deindustrialization" of this state and the New England region as a whole was predicted.

The "mature industries" turned out to be no longer competitive, so that Massachusetts seemed to have lost its industrial basis. The U.S. Department of Commerce predicted that the standard of living would fall for the once so rich New England below the national average - and continue to fall. A few figures elucidate this tendency: In 1975 unemployment in Massachusetts reached its highest level with 11.2% which was almost 3% higher than the national average (8.5%). A look at the statistics of the employed is even more significant. In the first half of the 1970s the rate of employed in non-agricultural fields shrunk by 0.4%, while it grew in the national average by 9.9% at the same time. Similar trends can be shown for growth and income development. In 1982 the average income was still 10% below the national average.

Even though Baden-Württemberg was at no time as low, compared with the national average, as Massachusetts, it was considered to be especially susceptible to fluctuations of raw materials prices, due to the lack of own resources coupled with a high percentage of traditional industries. After an enormous cut-down of jobs in the early 1970s the labour market situation at the end of the decade was marked by stagnation in the number of jobs on a low level while the number of unemployed rose. Long-term structural trends were pointed to which would have negative effects especially for Baden-Würtemberg: Labour-saving technological progress would cause redundancies in excess of the number of jobs created through the build-up of capital and increased demand from the service sector. Also rationalization measures in the service sector would lead to a further reduction of jobs so the service sector would no longer serve as a buffer for unemployment. Furthermore, the high international differentials in social service and wage costs were said to be responsible for increasing difficulties in the markets. (See report of the Battelle Institute,"The Labour Market in Baden-Württemberg" of 1978.) These dangers were considered especially grave for Baden-Württemberg because this state is extremely dependent on manufacturing.

The steel, machine, and automobile industries are the biggest employers in Baden-Württemberg. In 1960 this sector employed about 10%, and in 1975 already 14% of the work force. Of these industries machine construction is the most important. About 263,000 jobs were provided in about 1,600 industrial plants in 1975, which is more than 59% of all jobs in the steel, machine, and automobile industries. The special problems for these industries and, therefore, for the state of Baden-Württemberg were, next to the trends already mentioned, their vulnerability to rationalization and, therefore, a loss of jobs, especially as seen in the heavy dependency on exporting by the manufacturing sector. The future perspectives were said to be especially gloomy in face of the new competitors from the so-called newly industrialized countries producing with much lower costs.

The Emilia Romagna, finally, seemed to have better conditions for a later boom. The unemployment rate at 5%-6% in the 1970s was up to 2.5% below the national average. In addition the growth rates were higher than those of the nation as a whole. But the Emilia Romagna also had some special conditions that were considered disadvantageous in comparison with the national average. For example, the share of employees in small companies was a lot higher than the national average. Of those employed in manufacturing 27.8% were in small firms with nine employees or less; far more than 50% worked in companies with less than 50 employees. In the 1970s the share of firms with less than five employees was higher than 80%. Next to a strong agricultural sector most workers were employed in the problem sectors of textile, clothing, and footwear and machine construction and metal industries. Thus it is not surprising that an analysis by economists, dating from the late 1970s, came to the result that due to the nearly exclusive organization of production in small and smallest firms which could not profit from returns on scale on mass production, the future prospects for the Emilia Romagna could not at all be described as bright. In the long run, price competition with other economic regions would work against her.

3 The Economic Boom

In the 1980s these pessimistic prognoses had to be revised completely. All three regions prospered. Growth rates, employment figures and income were consistently above average, while unemployment stayed far below. There is full employment in all three regions.

However, despite the similarities in success the underlying growth models differ. In Massachusetts there has been a basic change in the economic structure. The industries which once dominated have nearly vanished. High-tech industries and more lately services have taken their place. In 1975 about 23,000 jobs were attached to the high-tech field; in 1985 this figure has risen to 60,000. Thus, the share of high-tech jobs is above 10%. These jobs are in 900 businesses. Apart from a number of big businesses, small firms predominate. For an estimation of the total employment effect, the high spillover effects of high-tech businesses on other industries have to be considered.

In Baden-Württemberg the "old" industries have not only survived, they have even become the motor of growth. In this field there are labour-intensive manufacturing industries, mechanical engineering, textile, print and paper as well as precision engineering firms. The automobile industry is represented by the biggest German company and the biggest parts supplier in the world. Between 1970 and 1974 employment in the automobile industry rose by 20%, all together there are 264,000 employees. The manufacturing sector is extremely well developed with a share of 52% of jobs and an almost 55% share of the gross national product. The handicraft sector employs 704,000 persons in 99,000 businesses. With regard to the distribution of business sizes figures reflect a concentration of medium-sized businesses. There are 50 large businesses with more than 2,500 employees, only 9% of firms have 20 to 49 employees, and the biggest share of all firms is to be found having 100 to 200 employees. The success of these firms and of the region as a whole is largely accounted for by exports; the products are considered to be of high quality.

In the Emilia Romagna small firms have not disappeared, rather a new "founder wave" has been triggered. At present there are about 130,000 handicraft businesses and nearly 6,500 businesses in the manufacturing sector in the Emilia Romagna. Dominating are machine engineering, automobile construction, metal processing, textiles and ceramics. No business has more than 4,000 employees, 20 have between 1,000 and 4,000 employees, 96% have less than 250. The biggest share lies with the smallest firms. Especially handicrafts have attained new prosperity.

4 Attempts at Explanation

The reasons for these very different development processes are partially deeply rooted in the historic traditions of the regions. In Baden-Württemberg, for example, the development of industry and handicrafts is closely related to the former inheritance laws. The so called real partition causes an increasing splintering of real estate. Therefore, farmers were forced to give up farming or to earn extra money in industry or handicrafts. Even today the additional work in industry by farmers is an evident employment pattern. It has, for example, positive effects on the stability of employment and leads to a high identification of employees with their firm and, therefore, also positive effects on the quality of their work.

The Emilia Romagna, placed in the plain of the Po River, is a region that is favoured in respect to climate and traffic. It has both an agrarian and an urban tradition. Thus, the process of industrialization and mechanization is inseparably connected to agriculture, and the development of handicrafts and science with the prosperity of cities and trade.

For the many small firms the regional demand is still an important factor. In recent history the structure of small businesses has been favoured by the company policies of "productive decentralization" as an answer to the increased power of large scale businesses. The small firms won subcontracts from big industrial enterprises, which they could honour less expensively due to the absence of collective wage agreements. Only in the 1980s was the dependency of small firms on the large-scale businesses loosened. Also significantly responsible were the innovations in the field of information technology, from which especially the small firms benefitted, as they did from the changes in the markets. The strategy of "flexible specialization", which means supplying the smallest markets with high-quality products has proved, decisively advantageous compared with inflexible mass productions.

Historically an immigration region directly on the Atlantic coast, Massachusetts with connections to the industrial areas of the Mid-West originally led in the process of industrialization. Until the mid-1970s the traditional industries, textiles, machine engineering and steel dominated. Profits from these directly or indirectly financed science there. Today industry still forms the basis of Massachusetts' infrastructure.

The recent success of all these regions relies to a large extent on political patterns, that in spite of major differences do show some strong similarities. Upon first glance the stability of the political system in all three regions is obvious. The regional governments have been in the hand of one party for years and even decades and in all three cases favour strong integration policies. Certainly basis for this is also the relatively high homogeneity of the population structure. The population of Massachusetts, for example, is 94% whites, 4% blacks and about 2.5% of Latin-American descent. The Emilia Romagna is, beyond the social barriers, unified by a common experience in the resistance against fascism. In Baden-Württemberg the industrialization process never succeeded in dominating the rural culture, allowing the integration of conservative, social-democratic and liberal proclivities.

Relative political stability fosters calculable political conditions for economic activities. Even though for the three regions the connection between political stability and the prosperous economic development can not be proven explicitly, there is some evidence for such a relation. All three regions show significant links between politics and the economy, despite the different party-political orientations of their governments. This can be seen in the economic policy measures for the improvement of the infrastructure. The education, communications, and traffic and social systems are exemplary in all three regions. The inclination towards consensus can also be seen in the relatively low wages during the declining phases of the business cycle and in crisis situations, without a drastic increase of strikes.

The three regions also show a certain abstinence from politics, which plays a great role for the founding of new businesses. The basic mood for start-ups is, at least in Massachusetts and in Baden-Württemberg, but recently also in Emilia Romagna, extremely positive. Business activities are being promoted rather than hindered. Furthermore, it is striking that industrial relations are not very contradictive in all three regions. The unions play no role in the case of Massachusetts. This also applies on the company level to the Emilia Romagna; even though the union movement has a long tradition there and is tightly connected with the political parties, it is barely represented in the small busineses. In Baden-Württemberg the unions and works councils, despite their strength, are participatively aligned.

Finally, a policy of actively guiding the industrial structure is being pursued in all three regions, as will be shown below. In Baden-Württemberg the state government is especially engaged in the field of education and vocational training. Based on the tradition of trade promotion, conceived in the nineteenth century by Ferdinand von Steinbeis, institutional conditions were created that enabled associations, guilds, handicraft chambers and other interest and professional organizations to create decentrally practice-related qualifications programmes. One important characteristic of this system of education and vocational training institutions covering the whole of Baden-Württemberg is their alignment towards small and medium-sized businesses, and they are supplemented through a network of colleges, universities and business advisory centres. Recently technology transfer centres, which create a link between basic research and practical application, were established in addition. The success of this policy can be seen in a, compared with the national statistics, above-average rate of skilled labour. Baden-Württemberg is also above the national average in numbers of engineers and scientists.

Educational and advisory measures are supplemented by an active infrastructural policy and an industrial location policy offensively carried through on the regional level. But special emphasis has been placed on technology promotion for small and medium-sized businesses which is basically founded on three programmes: (1) the promotion of development plans for small and medium-sized businesses (CI programme), which supports research and development programmes of small and medium-sized firms; (2) the programme for promotion of the use of modern technologies (MT programme) offers small and medium-sized businesses loans or subsidies for investments in innovative technologies and products; and (3) the "Promotion of the Founding of Technology-Oriented Business" (Förderung technologieorientierter Unternehmensgründungen = TOU) programme has decreased the financial risk of the starting up a new business with "pretentious technologies" by granting low-interest loans.

All together 2,122 businesses were financially supported from 1984 until 1988. The sum of the granted loans in the MT programme alone was 723,5 million DM; the sum of subsidies was 196 million DM. In the CI programme at the same time subsidies were 132,6 million DM. In the TOU programme the loans totaled nearly 50 million DM for 268 business foundings. Thus, the total output was more than 1.1 billion DM. Most of all the smallest size businesses with less than 20 employees were the beneficiaries of these programmes. The vast majority (more than 95%) of the promoted businesses were in the manufacturing sector. The focus was put on the promotion of machine construction and electronics. (See Becher et al. 1988)

Massachusetts has also pursued an active industrial policy since the late 1970s. It is based on support for industrial locations, qualifications programmes, creation of research institutions and conversion of so called mature industries. More than $ 3 billion in loans were negotiated in the field of venture financing with the help of the federal financing agency (Mifa). The Bay State Skill Corporation is engaged in the field of education and training. And last but not least there is the Industrial Services Programme that exists since 1984 that promotes the following institutions:

1. Workers Assistance Centers
2. Cooperative Regional Industrial Laboratories
3. Business and Financial Services
4. The Economic Stabilisation Trust

With the help of these organizations and the bundle of programmes instituted, efforts have been made to keep the traditional industries competitive and to train employees, who are affected by company close-downs, in the field they were qualified for. The most quoted example for a successful conversion of a city in decline, whose former strength was based on the textile sector, is Lowell. Only ten years ago the unemployment rate was at 13%, while now there is full employment with 2.9%. The biggest employer is Wang Computers. Of course, the establishment of the computer industry is also a result of the internationally unique infrastructure in the field of research. In Massachusetts alone there are 120 colleges and universities, providing an enormous reservoir of qualified personnel. The federal support for the growth of hard and software industries should not be underestimated. For example, nearly $ 7.7 billion in military spending is being transfered from Washington to Boston. It is estimated that this accounts for nearly 16% of all growth there. All together it must be stated that Massachusetts' boom is sponsored and promoted by the federal government.

It is not all that astonishing that the state plays a decisive role in the Emilia Romagna, which has been governed by the communist party uninterruptedly since the end of World War II. Development plans were worked out and so called comprensori, interfaces between the regional government and the 320 municipalities, were created. With the help of this instrument an active structural policy was carried out on the provincial and municipal levels, concentrating on an improvement of the infrastructure and the educational measures. Rather more astonishing is the fact that the private initiative of entrepreneurs is promoted in combination with a decentralization of decision levels with regard to particular industrial structural measures. An important element is the preservation of mature structures. This concerns most of all the promotion of handicrafts and support for the start up of small and medium-sized businesses.

Since 1974 the small firms are being supported by the regional development agency ERVET. Its role in terms of the philosophy of the regional policy is paradigmatic. The parties involved in ERVET are the regional government with a 51% interest and associations, individual firms and banks with 49%. ERVET promotes research and technology consultation and has created information centres for single industries such as engine building, ceramics, footwear, metal-mechanical manufactures, and textiles, as well as for construction and farming. Industry-overlapping, so called horizontal projects, have been established, which concentrate on training, management, ecological advisory and financing. The idea of planning, established in the socialist-communist tradition, was left behind for the sake of a decentral model of "controlled growth", so the programme of ERVET. There is no long-term orientation at planning data, but rather a supply of business-specific services. There is also no regionally oriented economic promotion policy in the traditional sense, but rather a so called territorial policy. The functional units production, technology development, economy, finance, environment and science are integrated for the territory. The industry-specific information centres are organized accordingly along the same pattern. The way these work is best described using the example of CITER, the information centre for the textile industry.

CITER was founded in 1980 and meanwhile has 500 affiliated firms. It has a turnover of 1.8 billion Lira. CITER supplies its members with information, which enables them to perceive trends in fashion, production techniques, the international market, and society and consumer behaviour early enough to adapt to the conditions. Its region or industry specific function is, stated roughly, the provision of strategic information for small, independent companies. The technical, but also the social science faculties of the universities are included in this process. Economists at the University of Modena constantly analyse the competitiveness of the textile firms connected through CITER in comparison with international competitors, either large companies or other regions.

5 Similarities of the Development Models

The following factors come into play when the patterns of success of the three regions are compared: The three regions examined apparently have in common their great number of specialized plants and businesses, whose horizontal and vertical connections are mediated through state policy. Small and medium-sized businesses as well as handicraft establishments play an important role in all three regions. Therefore, entrepreneurial initiative plays an important role for the founding of new businesses but also for innovation-oriented company policies. In all three regions, however, institutions are established that offer company-relevant, production-related services. Economic policy is conceived to have indirect effects, through infrastructural policy, advisory institutions and indirect technology promotion. This policy is strongly oriented to the needs of the existing businesses. In all three regions special significance is attributed to infrastructural measures which include an explicit education policy, but also the territorial shaping of economic domains and the cooperation with existing institutions and interest groups. The general policy of all three regions is also aimed at a social integration of its residents. Social integration here means the decentralization of economic decisions, the fostering of social mobility, the fight against disadvantages and the integration of interest groups into the general economic policy. In all three examples the utilization and disclosure of what has been called endogenous development potentials has been achieved by integrating different social functions.

Certainly the objection can be made that structures already existing in the examined regions favoured the accomplished restructuring and that regions with other economic characteristics would not be able to copy this positive development. However, the comparison of prospering economic regions with such different conditions shows that there is not one single way to success, but rather a basic pattern for the construction of an institutional framework for the development of specific economic activity.

The argument of uniqueness of conditions also has to be seen in a relative way. The basic pattern of each successful policy is the promotion of endogenous resources through the interconnection and combination of territorial, social and functional levels. This does not happen through direct political influences, but rather through exertion of influence on the general conditions. The state does not appear as the "omniscient leader and manager" of the economic processes, but rather concentrates on the management of the basic conditions for economic activity through well-aimed policy.

6 The Dangers of Success

Each of the three regions' development models also carry specific dangers that can already be discerned and to which the policies must react. The greatest threat to Massachusetts would appear to lie in the mono-orientation towards computer development. When the exorbitant growth potentials are ladled out, the problems in this field will assimilate with those of other industries. A first taste of future problems was the wave of dismissals in the computer sector in 1986. The reduction of jobs in the computer industry (at Wang Computers and General Equipment alone about 6,000 employees were laid off) could still be compensated by other pretentious embryonal high-tech industries. Nowhere else is there such a concentration of businesses in the fields of optoelectronics, genetic engineering and photovoltanics. In any case, for the time "after the computers", how research can be transformed to products will be decisive. Even today the comparative advantages of Massachusetts in the construction of computers and the testing of pharmaceutical products seem to be vanishing. Labour-intensive undertakings are threatening to leave the region.

The Baden-Württemberg model also is developing endogenous problems. The tight network of specialized small and large businesses has triggered a "deglomeration" process, leading to enormous ecological, social and infrastructural problems. For many businesses it has already become more profitable to invest in less developed regions. The strong orientation to the world market holds additional dangers.

Whether the development model of the Emilia Romagna can be continued without major ruptures is also questionable. The future changes in international competition in connection with the EEC integration would not appear to be irrelevant to the development of this region. Even today the level of investment required for retaining competitiveness surmounts the financial means of many small businesses. Company mergers seem to be recommendable, but this would probably lead to a loss with regard to the dynamics of many small firms.

For a solution to these problems, and this should not only be important for the regions examined, Massachusetts, Baden-Württemberg and the Emilia Romagna could learn from one another. Massachusetts offers an example for successful transfer of knowledge from the universities and basic research institutions into businesses. Baden-Württemberg is a good example for the transfer of high-tech into traditional products and production methods, while the Emilia Romagna is a role model for a flexible economy that is based on small and medium-sized businesses.

Against the background of the above arguments, the following theses can be put forward that go beyond a direct comparison of regions and can serve as a basis for a successful regional policy:

1. Important to the success of regional development projects is a stable political system, its social relations and the social integration of its employees and employers. A clear political vision, based on the traditions of the region, has a stimulating and motivating effect on the participating social groups. The improvement of the infrastructure through public policy is then especially significant.
2. Protectionist preservation strategies in mono-structured industrial regions are just as wrong as active policies that are based merely on technology parks and innovation centres, in other words, trying to copy the Silicon Valley model. Policies are successful when they influence the institutionalized conditions for economic growth and make use of the endogenous development potentials. Hereby the competitiveness of the traditional industries should be strongly promoted .
3. However, endogenous development potentials have to be made accessible permanently and transformed to real development processes. This transformation can only be carried out from the bottom up. Obviously a decisive factor is the creation of a institutional network, which connects the politics, the economy and science horizontally and vertically.

References

Battelle Institute (1978)
Der Arbeitsmarkt in Baden–Württemberg. Unpublished report.

Becher, G., U. Kunze, O. Pfirrmann, G.H. Walter and E. Weibert (1988)
Zwischenbilanz der einzelbetrieblichen Technologieförderung für kleine und mittlere Unternehmen in Baden–Württemberg, Zwischenbericht an das Ministerium für Wirtschaft, Mittelstand und Technologie des Landes Baden–Württemberg, Projekt des Fraunhoferinstituts für Systemtechnik und Innovationsforschung (ISI) Projektnummer 105007. Unpublished manuscript, Karlsruhe.

5.5 Endogenous Development Strategies in Agglomerations

Jürgen Allesch

1 Endogenous Strategies and Agglomerations

The agglomerations of Europe are undergoing a process of structural change (see Ewers et.al.1986). Having mainly developed as centres of industry and manufacturing they have come under strong pressure to restructure as traditional industries have fallen into decline and the creation of wealth has become increasingly dependent on factors such as science, creativity and qualifications rather than classical infrastructure and natural resources. These restructuring processes are extremely successful in some agglomerations, in others they are long lasting and painful. Those agglomerations with a strong position in research and development (R&D) and in services industries tend to be successful while those lacking these factors are in danger of being left behind.

In this atmosphere of economic change those regional policies have proved successful which aim at the development of the endogenous potential of a region, thus, its existing structure of businesses and skills. Agglomerations, with research institutes, universities and colleges, as well as private and public service institutions often already in existence, are in many cases well suited for policies aimed at strengthening the capability of local businesses, especially small and medium-sized enterprise (SMEs) to invent and innovate, thereby developing the region's endogenous potential to compete successfully in new markets. In what follows the main instruments of an innovation oriented endogenous development strategy for agglomerations will be explained, using the case of Berlin as an example.

2 Instruments of an Innovation Oriented Endogenous Development Strategy

The characteristic principles of endogenous strategies in regional policy can be defined as follows:

1. Endogenous strategies in regional policy are applied specifically to the strengths and weaknesses of a region. Regional economic policy concentrates, therefore, on maintaining and promoting existing businesses.
2. Endogenous strategies for agglomerations are technology- and innovation-oriented. The promotion of small and medium-sized firms and of new businesses, as well as the improvement of technology-oriented cooperation, e.g. between universities and business, receive high priority.
3. The promotion and further development of an existing industrial structure, i.e. of an endogenous potential based on smaller and medium-sized businesses differs considerably from the regional economic policy of the 1970s. Under the present economic conditions a considerable part of the measures for the promotion of technical and marketing changes in small and medium-sized businesses are concerned with real transfers (i.e. technology transfer, vocational training, consulting, information and cooperation). These measures are tailored to the particular needs of the local economy; they also take into account the special possibilities provided by the infrastructure and the locally available resources of research and development.

Endogenous policies, thus, supplement financial support for the regional economy with real transfers (see Kern in this volume). The following representation of the different operational levels of a regional strategy of technology transfer and innovation consulting (see Fig. 1) draws on more than ten years of experience made in Berlin in the process of developing a differentiated technology transfer landscape. Its fields of operation may be transferred to other regions as well. In practise these policy elements may be either centered in one agency or spread over various agencies in cooperation with each other.

2.1 Operating Levels of Technology and Innovation Oriented Real Transfers

The support of SMEs in their efforts to master structural change is an important element in the improvement of the competitiveness of an agglomeration. The intensification of innovation consulting and technology transfer from universities and research institutes to SMEs concentrates on the areas of:

1. Technology transfer
2. Personnel transfer
3. Cooperation
4. Start-up consulting

Figure 1: Regional Supporting of Innovation by Real Transfers

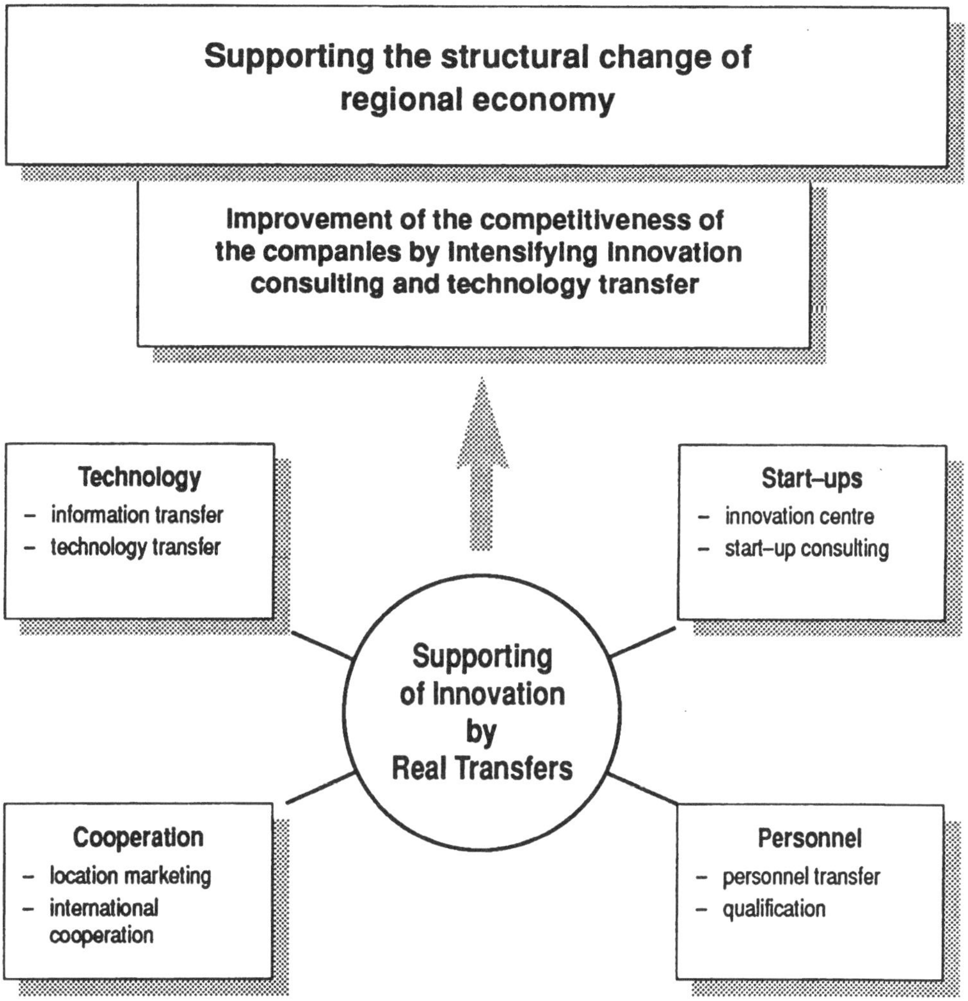

These fields of activity will now be shortly discussed.

2.1.1 Technology Transfer

The operational level "technology" may be split into the areas of "technology transfer" and "information transfer". Technology transfer combines direct consulting and contact mediation. Qualified consulting for all areas of technology can not be provided by the personnel resources of a single unit with regard to all technological details. Consulting service consists, therefore, in mediating between business on the demand side and experts who can supply suitable solutions. The mediation service also comprises medium- and long-term project counselling, contract research, licenses and research cooperation. For the mediation it is not necessary that the experts be from within the region. Of basic importance is the ability of the mediator to establish and to serve cooperation and information contacts.

Information transfer, i.e. the obtaining, distribution and processing of information, is gaining increasing importance in the decision-making process of businesses and for market success. The increasing amount of information coupled with its rising complexity corresponds to new technical possibilities for technology transfer. Computer-based data banks are becoming ever more the technical nucleus of modern data transfer and processing. The possibilities for market, technology, patent and licensing enquiries have only just been tapped. The 3,000 commercial data bases distributed worldwide are - particularly by small and medium-sized businesses - not being used to their full potential. It is an important task of regional innovation promotion to ensure that the international supply of information is transfered to regional industry.

2.1.2 Personnel Transfer

The establishment of qualified personnel in the businesses of a region is decisive for its stability and the further development of its competitive ability (see Martin/Röseler 1988). The area of personnel or "manpower policy" may be divided into personnel transfer and qualifications. Personnel transfer aims at establishing qualified personnel resources in SMEs. The "transfer of heads" i.e. the mediation of the latest technological and business management knowledge by placing qualified young business professionals and engineers in SMEs is, from the individual business point of view, one of the most effective forms of technology transfer.

Qualification and further education for employees of SMEs aid the firm's integration of the technological transition. The success of a technology depends upon its users' ability to apply it. This means that the employee's individual characteristics of qualification and performance have to correspond with the demands of the firm and the respective technology. Further education of employees has become a prerequisite for the future competitiveness and survival of every business firm.

2.1.3 Cooperation

The significance of cooperation for agglomerations needs no further explanation, especially in view of the future European single market. Cooperation can be split into location marketing and international cooperation. Location marketing comprises the compilation, processing and supraregional presentation of products, processes and services offered by or produced in a region or on location. Brochures, exhibitions, fairs and congresses, and - as described above - data bases are the instruments for spreading awareness of the offers and advantages of a region beyond the region itself.

Supraregional marketing is particularly difficult for SMEs and their products and services. Regional data banks which market supraregionally the services offered by businesses and research agencies will in future take on greater significance in national, European and international competition. The compilation, preparation, and supraregional distribution of regional services in technology, products and services by technology transfer units need to be utilized more as the advantageous instrument for location marketing that they are.

2.1.4 Start-Up Consulting

New technology based businesses emerging from already existing ones or from research institutes are an important factor in overcoming the difficulties of industrial development. The promotion of start-ups can be divided into the areas of start-up consulting and technology parks.

Start-up consulting includes, above all, consulting with regard to product ideas and technical and market conditions as well as financing and promotion consulting. Particularly technology-oriented founders - for example, university graduates, with their technology-oriented scientific backgrounds - need this kind of consulting to succeed with their technological product idea. It must be noted that new business consulting need not be limited to technology, particularly for peripheral regions.

Technology parks (or innovation centres or "incubators") have experienced an almost saturating distribution throughout many European countries (see Allesch (ed.) 1989). The significance of technology parks as economic-political structural measures lies primarily in their "thrusting function", in other words, in the establishment of an economic and socially positive climate for technology-orientated start-ups. Their success is manifested in the individual start-ups of a region.

The integration of the fields of activity described may be summarized as depicted in Fig. 2 using the Berlin experience as an example.

Figure 2: The Berlin Model

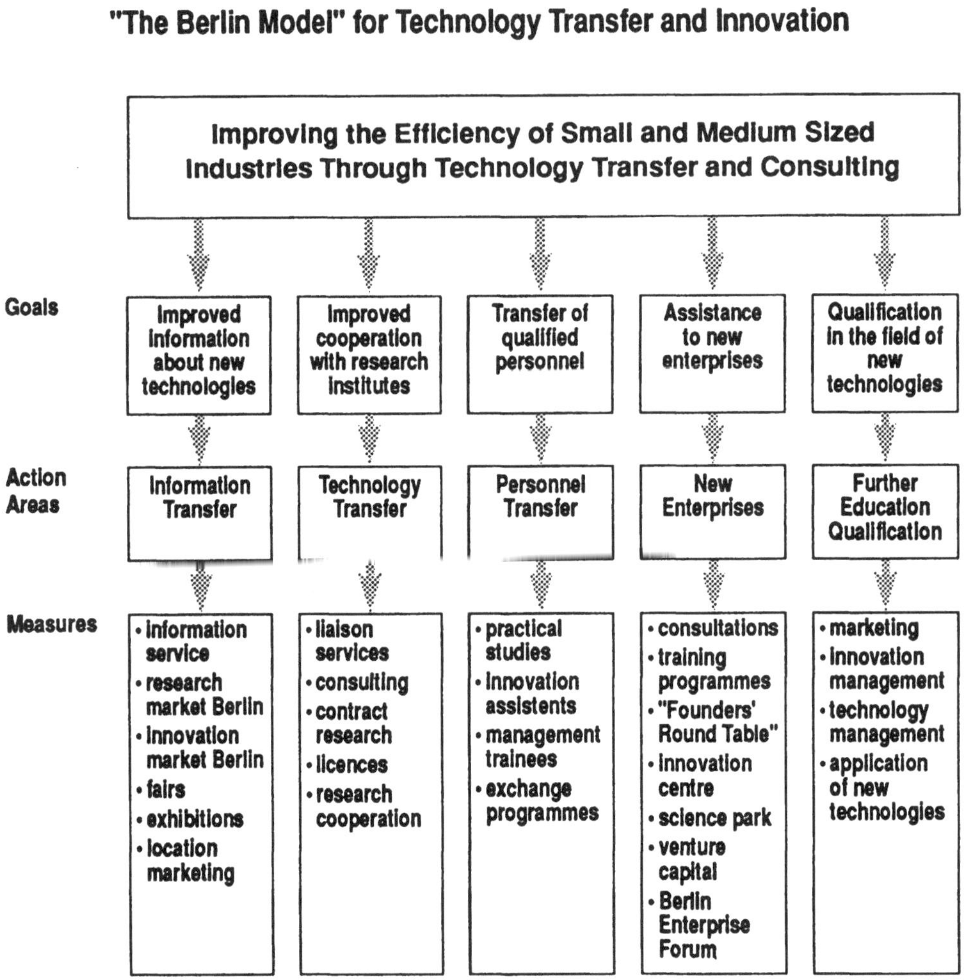

3 Building Endogenous High-Tech Industries in the Framework of Innovation-Oriented Regional Development - the "Berlin Model"

3.1 The Political and Economic Situation of Berlin

The city of Berlin is closely integrated politically, economically and socially into the Federal Republic of Germany (FRG) (see Kern 1986). Despite its peculiar geopolitical location its economy is comparable to the rest of the Federal Republic of Germany in efficency and dynamics. Berlin is still the largest industrial city in Germany in terms of numbers of companies and second after Munich in numbers of employees. Berlin's annual per capita gross national product is above the average of FRG, its total GNP was in 1985 one-half of that of the smaller Scandinavian countries.

Nevertheless, on closer inspection a number of structural deficits of Berlin become obvious. The loss of its function as the German capital, which accounted directly and indirectly for the biggest share of the city's revenue before World War II, meant that since 1945 Berlin has had to make its living from industry alone, which has not been made easier by the continuous loss of headquarter functions, banking, and insurance during the Cold War period. Moreover, industry itself in Berlin suffered from structural problems having been mainly oriented towards the major investment goods branches of industry, such as electrical and machine engineering, which have been undergoing major structural changes in the course of the micro-electronic revolution. This restructuring coincided with a period of general slowdown of economic growth and during the 1970s led to substantial losses in employment in Berlin's industry. In this situation the Berlin Senate in the late 1970s adopted a comprehensive strategy to overcome the structural deficits of the Berlin economy. The aims of this programme which later served as a model for other German regions were:

1. To modernize the industrial structure of the Berlin economy
2. To increase the competetiveness of firms
3. To improve the skills of both staff and management
4. To foster new companies, especially in high-tech areas
5. Most important, to create as many new jobs as possible in all branches of industry and the services sectors with expanding markets

In doing so the Berlin Senate aimed at developing the endogenous potential of the city which is especially to be found in its high-profile in science and technology.

Figure 3: Technology Transfer in Berlin

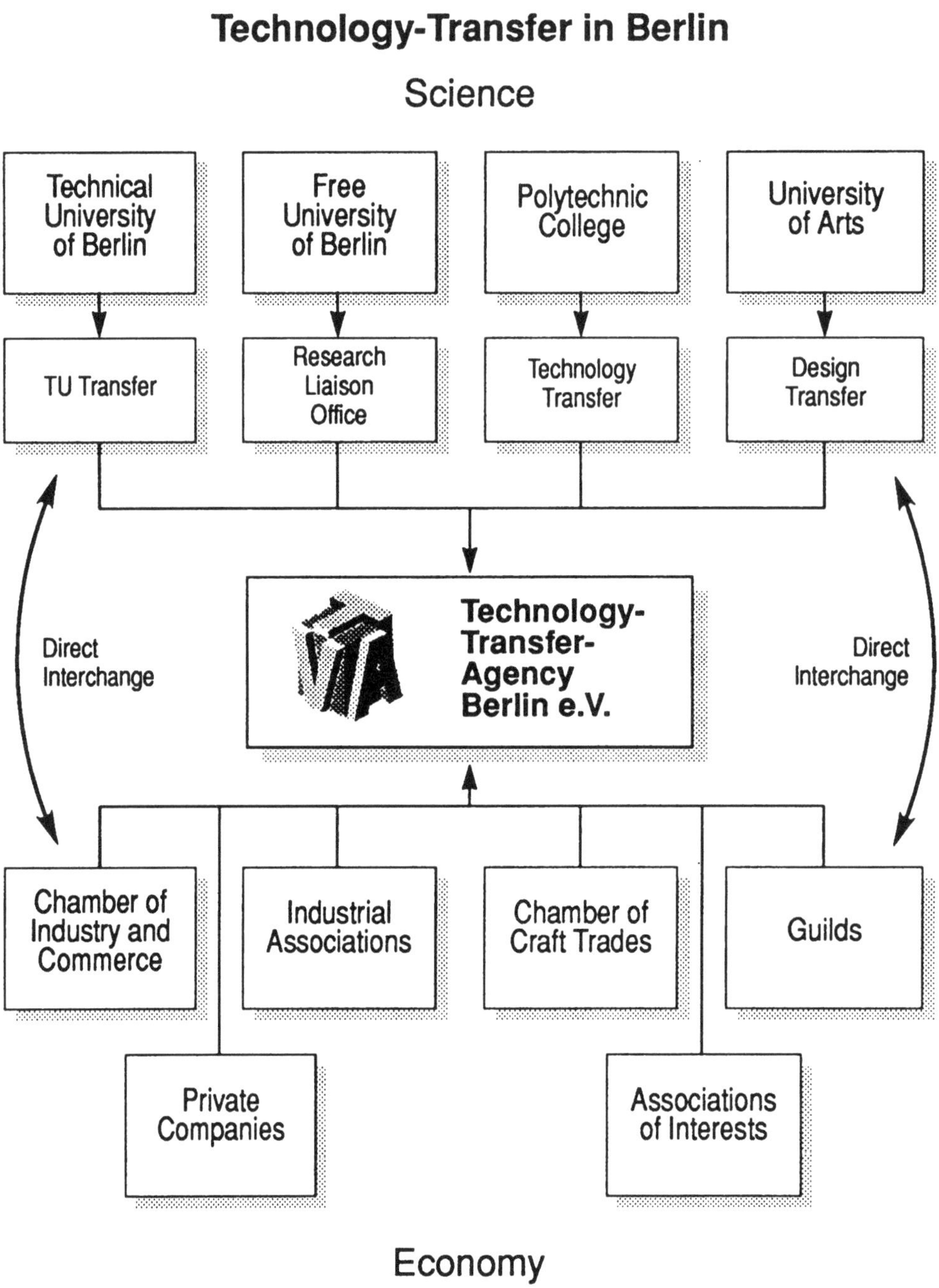

Berlin is home to two large universities and several colleges of higher education with a total of 95,000 students. There are about 180 public and private research institutes and all in all 40,000 employees are working in R&D. This potential in science and technology makes Berlin a major centre of research and development in Europe. In 1983 the "Berlin Centre for Innovation and New Enterprises" (BIG) was founded as the first German innovation centre, followed by the "Technology and Innovation Park Berlin" (TIP) in 1985, both of which strengthened the city's position in the application of future-oriented technologies by triggering an increasing number of technology oriented start-ups of new companies.[1]

This development was supported by several technology-transfer agencies which support the adaptation of new technologies within traditional industries. The most important of them are the transfer agencies of the Technical University, of the Free University and of the Technical College; the consulting services of the Chamber of Industry and Commerce, the technology centre of the association of German engineers (VDI/VDE), and the Technologie-Vermittlungs-Agentur Berlin e.V. (Technology-Transfer-Agency, TVA) which functions as a central mediator between science and regional industries (Fig. 3).

TVA is an institution situated outside the university and supported by industry, industrial associations, the Berlin Senate and science and research institutions. More than 160 small and medium-sized enterprises (SMEs) are registered as members of TVA. A specific feature is its close connection with the Berlin Senate's innovation oriented policy of regional development in which TVA has the vital role of translating policy objectives into practise. In its ten years of operation an approach to advisory services has been developed to help SMEs with all the different kinds of problems that they are faced with in the course of realizing their strategies for future success. Thus, working as a "one-stop agency" advice is offered on the basis of the synergetic fusion of TVA's main branches of service, i.e. information service, technology transfer, personnel transfer and advice on environmental issues in a form applied to the specific problems of SMEs.

---

1 While this paper went into press, the opening of the East German border changed the medium term prospects of the Berlin economy considerably. However, all future developments will take the structures as described here as starting point. The innovation-oriented endogenous policy recommended in this paper will be even more important in face of the challenges of the future.

Figure 4: A Holistic Transfer Model

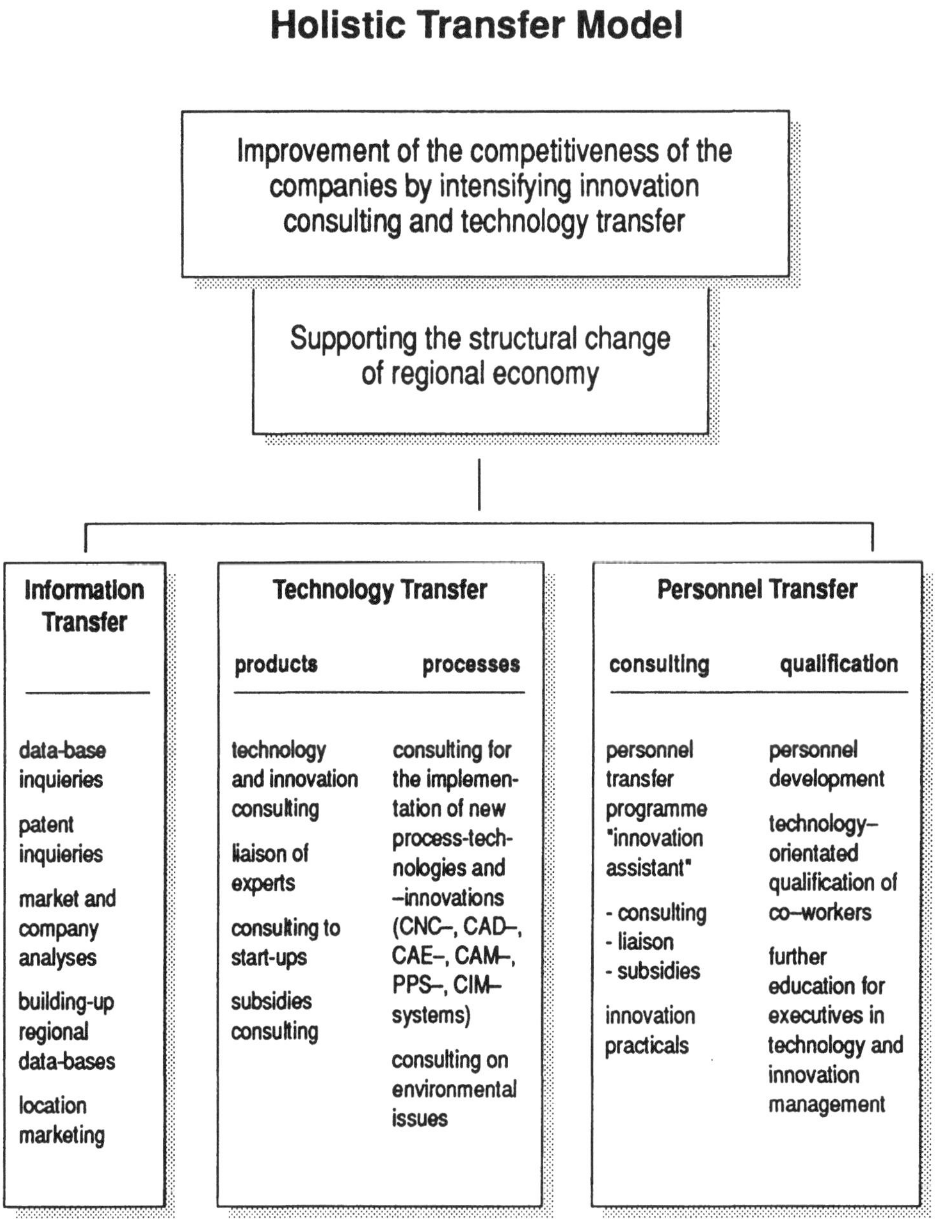

3.2 The Consultancy Services of TVA Berlin

In the following a brief description is given of the different operational levels of the "Berlin Model" as carried out by TVA in cooperation with other Berlin agents in technology transfer (Fig. 4).

3.2.1 Information Transfer

As pointed out above, information transfer is one of the central tasks in the transfer of scientific knowledge. The TVA information service provides support to management through:

1. An extensive online data bank service with access to 900 national and international data banks, that means to state of the art science and research
2. Inquiries about market situations and technical standards

To date, TVA has answered more than 3,000 data bank inquiries. The importance and value of information as the basis of management decisions has been increasingly recognized in recent years, leading to a growing number of inquiries to TVA as regional information broker. Recent experience shows that SMEs are prepared to carry some of the costs of information requests especially as so-called "value added services" are concerned, where the innovation consultant surveys and interprets the data bank information provided. Thus, information services are integrated into the general services of technology transfer and innovation consulting.

3.2.2 Technology and Innovation Consulting

SMEs often lack the know-how necessary for their own research and development without outside help. They lack:

1. Personnel and technical expertise
2. Financial resources
3. Contacts and experience with external opportunities to use new technology

TVA alleviates these limitations through its advisers who mediate between SMEs on the one hand and outside experts on the other. A problem analysis is provided, appropriate partners are searched for and first contacts are arranged. Help is provided to make draft contracts and give support to cooperative projects. Furthermore, TVA provides information about public financial aid programmes. TVA has published the "Förderatlas Berlin" a brochure which gives an overview about all public grants-in-aid programmes existing. Also part of TVA's technology and innovation consulting service is to assist new technology-oriented enterprises in getting through the difficult start-up phase. The range of services for start-up companies includes:

1. Evaluation of the innovation concept with regard to technology and marketing potential
2. Arranging for expert advice on special problems
3. Financial couselling public funding, venture capital)
4. Assistance in recruiting staff members

3.2.3 Production Technology Consulting

The technology and innovation consultancy service of TVA not only includes product innovation, but process innovation as well. TVA supports the efforts of Berlin's SMEs concerning the implementation of modern production technologies. More than 100 companies have received comprehensive consultancy services in modern production technology in recent years.

As a new service TVA offers consulting on environmental issues to SMEs. This programme aims at supporting SMEs in solving environmental protection problems arising in connection with their production activities. The programme was started so recently, that no experiences can be reported.

Figure 5: Personnel Transfer Programme "Innovation Assistant" Number of Requests and Recruitments

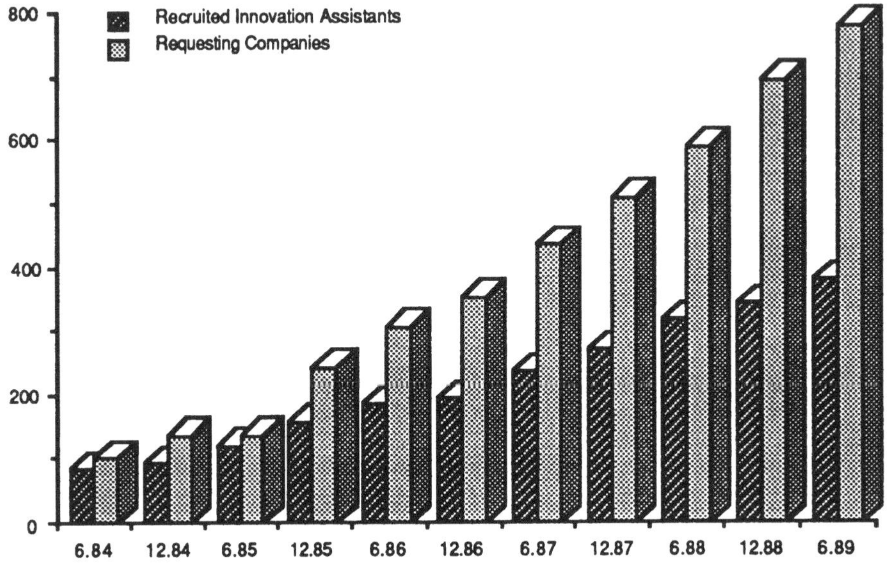

3.2.4 Transfer of Personnel

A well-qualified staff is the prerequisite for innovation and for dynamic business development (see also Kern in this volume). TVA's personnel transfer service assists Berlin firms in recruiting qualified university graduates and postgraduates as specialists and management trainees (innovation assistants). As recruitments are always risky, the Berlin Senat Department for the Economy sponsors the employment of innovation assistants by covering 40% of their taxable gross salary for a period of 12 months, the maximum sum being DM 24,000. Only those personnel service businesses and manufacturing companies are eligible for support whose annual turnover does not exceed DM 50 million per annum. The aims of the measure are: firstly, to facilitate the recruiting of university graduates in SMEs, which is also intended to reduce prejudice against university graduates which tends to exist in SMEs; secondly, to open up new occupational fields for university graduates, thus, to facilitate a reorientation of university graduates towards SMEs as a possible source of employment. An overview of the development of the programme "innovation assistant" is given in Figs. 5 and 6.

Since the start of the programme in September 1982 more than 500 companies have been advised on the programme and more than 800 innovation assistants have been

Figure 6: Personnel Transfer Programme "Innovation Assistant" Companies Recruiting Innovation Assistants by Number of Employees

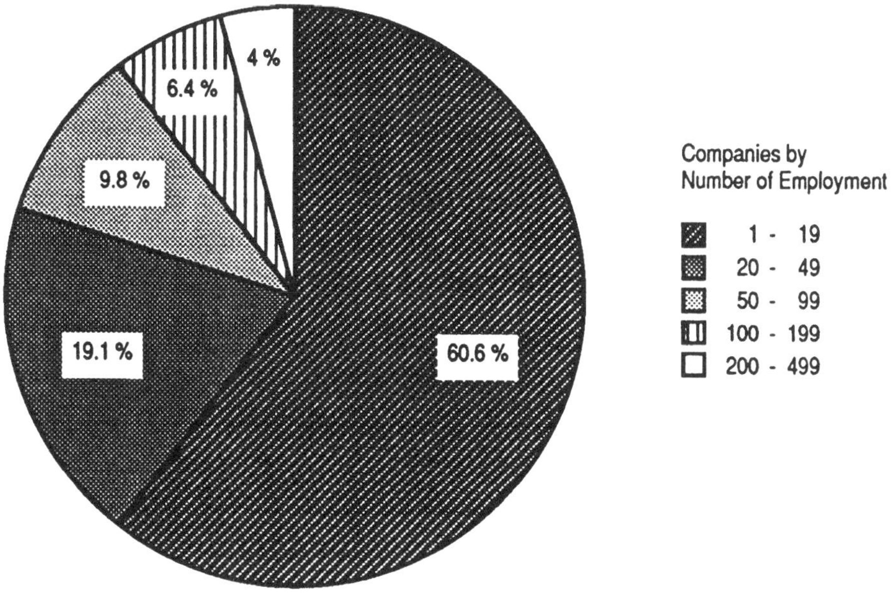

placed within Berlin SMEs. There is a great demand to be met in micro electronics, information and data processing. About 60% of the companies supported had less than 100 employees; more than 50% of the firms had no prior experience with university graduates. The central importance of the personnel transfer programme lies in the field of innovation support. The transfer of new and qualified know-how enables SMEs to start new programmes and to extend R&D activities.

The experience with the personnel transfer programmes shows that they are tailormade for the requirements of SMEs who use this service extensively. As a next step, the quality of the service is going to be improved by offering special training programmes to prepare "innovation assistants" for the special task of working in SMEs. Such "training on the job" programmes will cover environmental protection techniques, marketing for the European single market and international cooperations, quality control technology transfer, and management skills. These additional qualifications areas are intended to prepare young staff members for their specific management tasks.

3.2.5 Summary

Summing up, it may be noted that not a single measure but rather a combination of measures at different operational levels constitutes the "Berlin Model". Its practice has made obvious the fact that the main factor of a successful regional technology transfer scheme is the availability and recruitment of qualified personnel. Thus, the formerly merely technology-oriented/supply-oriented approach was extended into a holistic demand-oriented approach covering a wide range of advisory services taking the form of a "one-stop agency" for different regional promotion programmes. Such an agency also serves as a public-private interface dealing with questions and programmes to promote the region's competitiveness.

4 The Creation of a Regional Innovation Network

The present situation of many agglomerations of Europe is characterized by a gap between the potential provided by innovation centres, high-tech companies and research institutions, on the one hand, and the traditional industrial potentials of a region on the other. Closing this gap is one of the vital tasks of an innovation-oriented regional policy. In Europe, transfer agencies are increasingly being included in concepts of innovation-oriented regional economic strategies. This raises the question of how effectively can technology transfer and innovation consulting realistically be used as an instrument of regional economic policy. At the heart of the "Berlin Model" of technology transfer and innovation consulting is the idea of taking the potential of existing industrial and service firms as the starting point for an innovation-oriented strategy.

Starting from the existing endogenous economic potential means that both, the strengths and weaknesses of a region must be taken into consideration. In contrast to traditional concepts of regional policy, which are usually directed towards attracting new firms, the support of existing SMEs receives a much greater significance. A policy concentrating on the strengths of a region is able to structure and optimize science and technology transfer. Technology-transfer agencies, thus, form an essential factor in creating and establishing the infrastructural framework to support innovation. Their importance for today's economy is comparable to that of canals and railways in the last century, or more recently of the telephone network and airports. The infrastructural factors governing the success of a regional policy oriented towards technology and development can be summarized under the concept of innovation networks, as developed by Bullock (1983) and extended here. The essential elements of a regional innovation network in a so-called high-tech cluster are shown in Fig.7.

The infrastructural requirements can be divided into the fields of politics, transport, supply and demand, labour market, training and qualifications, R&D potential, and social environment.

With the framework of such mutually influencing aspects of a regional innovation network, technology transfer and innovation-consulting agencies form part of a forward-looking regional economic policy. Clearly, transfer agencies cannot and should not attempt to replace the innovative efforts of SMEs. Rather their contribution should be to provide an infrastructure favouring innovation, enabling the individual firm to react more promptly and more comprehensively to changes in the markets and in technology. This means that in particular SMEs are given the help they need to be able to help themselves. From a medium and long-term perspective a varied, user-oriented apparatus of technology transfer and innovation consulting can help combine the realization of the potential of SMEs with the development of an innovative and thus forward-looking regional economy. An additional element with view to the long-term development perspectives will be the augmenting of existing regional and national transfer and innovation structures with an international network of information and innovation, as is presently being launched by the SPRINT programme of the Directorate General Regional Policy of the Commision of the European Communities.

Figure 7: High Tech Clusters

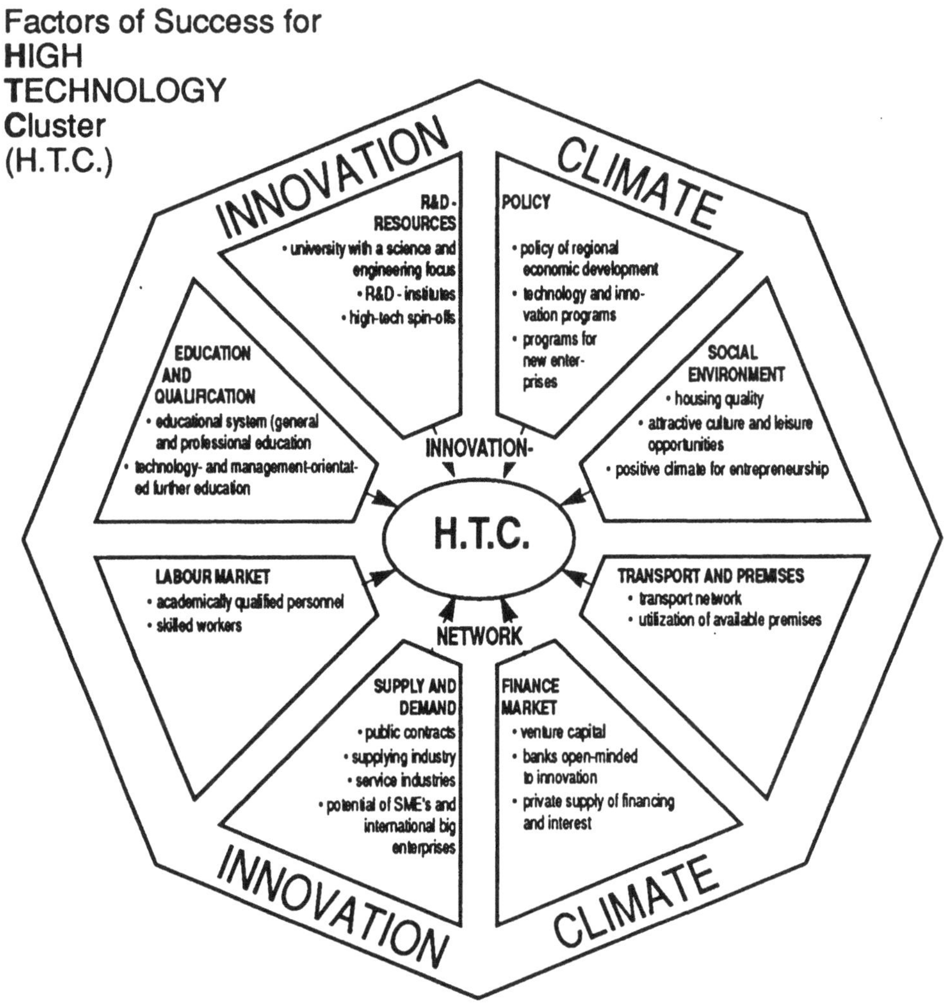

References

Allesch, J. (ed.) (1989)
Regional Development in Europe: Recent Initiatives and Experiences. Proceedings of the Fourth International Conference on Science Parks and Innovation Centres. Walter de Gruyter Berlin, New York.

Bullock, M. (1986)
Academic Enterprises, Industrial Innovation and the Development of High Technology Financing in the United States. mimeo London

Ewers, H.-J., J.B. Goddard and H. Matzerath (eds.) (1986)
The Future of the Metropolis: Berlin-London-Paris-New York. Walter de Gruyter Berlin, New York.

Kern, M. (1986)
Evaluation of Recent Policy Innovations in Berlin (West) in: Ewers, H.J. et. al. (eds.) (1986), The Future of the Metropolis: Berlin-London-Paris-New York. Walter de Gruyter Berlin, New York.

Martin, H.J. and V. Röseler (1988)
Innovationsassistenten: Personaltransfer als neues Modell der Innovationsförderung im Ländervergleich, Reihe Technologie Transfer, Bd. 13. Verlag TÜV Rheinland Köln.

6 Perspectives of Regional Development

6.1 Perspectives of Innovation-Oriented Regional Development Strategies

Walter Hirche

1 Preliminary Remarks on the Concept of Regional Development

When talking about "regional development", one must clarify the definition of "region", in particular because the terms have different meanings in the different languages of this Congress. For instance, in the invitation it was announced that a "Minister of a regional government", a "Ministre d'un gouvernement régional" would be speaking this morning. For a start it might be enough to say that regional policy and the economic policy of a Bundesland are not the same in my understanding.

The government of a state of the Federal Republic of Germany pursues regional development policy in two ways:

1. Within the borders of the state e.g. in promoting economic development for single regions or areas of the state or also in the states' border regions, sometimes in cooperation with the neighbouring state.
2. For the state as a whole, in the form of providing the basic structure of the economic policy.

Of course the issue is not only "Ordnungspolitik" but also specialized policy like energy policy, industrial policy, "Mittelstandspolitik" (small business policy) or technology policy. So the term "region" is ambiguous, it is determined more by socio-economical and cultural factors than by the borders of states or the country, even though, as a rule, identity of these aspects is given as a result of historic processes. Therefore, I see a region without stipulation as to a particular size, as being a geographical, economic and also cultural space, that distinguishes itself through a common will and capacity for organization. For this it is decisive firstly, that the capacity for organization does not necessarily derive from the political bodies and decision-makers although they unquestionably have an impulse and mediator function. Secondly, that the term "organization will and capacity" already contains the element of joint action.

Thus, at the moment, I do not have in mind the recent attempts to define Northern Germany as a region, but rather the regions which are for instance to be found within the state of Niedersachsen (Lower Saxony). The borders of these regions are determined by a topography of mutuality. This nebulous definition of a region, the dependency of concretizing on any particular situation, does not hinder us in pointing to innovation-oriented development strategies. The value and effectiveness of such development strategies, however, depend greatly on how they correspond with a certain geographically limited area and whether we will succeed in building up a common responsibility and organization for future strategies in these particular regions.

To that extent my apparently general preliminary remarks are already part of the subject. Regional development strategies, that is the message, can be shown as an abstract model only to a certain extent. Their usefulness depends entirely on whether we will succeed in grasping the status quo of a region i.e. its endogenous potentials and in developing concrete perspectives, based on these potentials, according to which the regions can support themselves. With regard to the future prospects of a regional development strategy only general points can be made. It is crucial that human and social factors of influence are recognized early. In this course the creation of an innovative climate in a region is more important than an exact addition of single instruments of development.

2 Some Instruments of Regional Development

It is a truism that the success of new instruments depends on how favourable the general conditions of the economy are for the development of a region. In this respect the competition of the different taxation and social systems plays an increasing role on the national level under the rubric of costs. On the state level in the Federal Republic of Germany the classical infrastructure and the stage of development of the modern infrastructure plays a major role. I can only refer to this point briefly here.

2.1 Classical Infrastructural Measures

As classically essential to the infrastructure of the economy are especially the prerequisites in the field of transportation i.e., convenient connections to a region by railways and roads, if possible also by waterway and air traffic. The existence of a sufficient number of regional airports as well as an early establishment of all telecommunication services in the regions are currently crucial points of transportation infrastructure. Sufficiently developed industrial areas are a further necessary part of the infrastructure. In addition, there are the following requirements for a modern infrastructure: The presence or proximity of important research centres, sufficient institutional possibilities for qualification (education and advanced training), as well as pollution abatement facilities for unavoidable or nonutilizable industrial waste of any kind.

2.2 Structural Policy

Structural change and the crises of certain industries, as can be seen from data concerning the labour market, lead to special problems when monostructures have existed over a longer period of time. Therefore, a permanent task for economic policy is not only to balance the disadvantages of location, but also to support the adjustment processes of the economy with regard to the necessity of structural change. For improvement of the economic structure it is important that growth-intensive branches of the industry are able to expand. Also attention has to be paid so that policy measures for regional promotion do not support an entrenchment of obsolete structures that hinder development. The mere preservation of jobs, as often demanded for the sake of a short-term political goal, can all the more drive a region into structural problems in the long run as can be seen, for example, in the steel industry. On the other hand, the growth of unemployment above a certain point is also a hindrance for comprehensive innovation.

Thus far the starting point for measures of structural policy has been the promotion of economic growth in structurally weak regions through immediate help for industrial enterprises. This also includes the support of investments for the improvement of the economy-related infrastructure. Both measures belong to the concept of "Gemeinschaftsaufgaben" (common goals) between the federal government and the states for the "improvement of the regional economic structure". For the fulfillment of these common goals the federal government takes part in the strategic planning and financing, while the rest of the financing and the execution of the regional economic policy measures are the duty of the state. But more and more it turns out that a policy of promoting economic development that only pursues the support of business expansion or only counts on grants for companies locating into the area, neglects the necessary modernization of the economy, its innovative orientation. Therefore, instruments of technology policy have been included in the promotion catalogue of regional economic development. But still this customary instrument of promotion is insufficient for a real stimulation of regional development, since an impulse effect is not a main part of it.

2.3 Research and Development Policies

Durable development in the regions will not be reached by copying the development in other places but rather through consistent reorientation that has to be innovation oriented. The key for a satisfying outlook for regional economic structures lies in the development and implementation of new technologies. That is why it is important to put research and technology policy in the service of economic policy, namely the regional economic policy.

Of course there is no sense in assigning research budgets according to a regional pattern, but when 2/3 to 9/10 of the economic growth can be accredited to technological progress, as economists tell us, then decisions in research and technology policy also have to be considered in view of their regional effects. Technological progress is in a way the leading member in the league of development strategies. No state in the Federal Republic of Germany and no region can manage without summoning up this energy. The development of a region or a part of a region has to suffer when it is systematically supplied with less means of promotion from the field of science and technology than the neighbouring region. This is why a vehement discussion has developed among the states on how a fair distribution of national funds for the advancement of research can be assured. For analysis has shown that an over-proportional share of federal means flows to only a few states in the southern part of the Federal Republic of Germany, which, of course, had a positive influence on the regional development and the regional labour market. Through this long-standing policy at the federal level, the historical economic structures have been stabilized considerably.

Let me offer one example: The construction engineering, electrical engineering as well as aviation and aerospace industries receive a significant share of the project promotion budget of the Federal Ministry of Science and Technology, because these industries are especially innovation oriented. These industrial groups, howevever, have a share of 30% of the overall economy in my state, Niedersachsen (Lower Saxony), 76% in Baden-Würtemberg and 81% in Bayern (Bavaria). Thus, the economic structure of the last two states, which is excellent at any rate, gets additional support with federal tax money, while the structural change and the innovation orientation in a structurally weak region remains underproportionately stimulated. Therefore, the regional policy of a state's government should also focus on a change within the national framework. This explains my demand for a new regional development strategy that is connected with the means for the promotion of research. What I have said for the states of the Federal Republic of Germany is in principle also applicable for the region of my state, Niedersachsen. There are strong interregional differences here as well and therefore, there is also the need for the introduction of new developments in the areas with weaker structures. But in nearly no case is it the single initiative that finally leads to success, but rather the sum of many coordinated steps that makes progress visible.

3 Promotion of Innovation

The more the term "innovation" has been narrowed down to the field of technology and even to products and processes, the more important appropriate facilities have become for regional development. Under the aspect of innovational dynamics, the location of technical universities, specialized institutions of higher learning and practice-oriented research facilities have special significance. They have such a high attractiveness for businesses and scientists, that fundamental imbalances between regions have resulted. As "fundamentally imbalanced" I consider those differences in the economy, technology and infrastructure that can not be balanced by the weaker regions or states themselves. Under such circumstances federal action is required. It is agreed that technology policy has to follow the principles of economy and competition in the free market economy. However, the priority of these principles has to end where there are disturbances to the structural balance in the sense of a loss of comparability of internal conditions of development and where the economy is too weak to stretch and rise by itself. In such a situation the government of the state of Niedersachsen, for example, has reacted with a technology programme that has an annual budget of 85 million DM. Let me mention just two of the measures, firstly economy-related research institutes and technology parks, and secondly, the promotion of technology transfer.

3.1 Research Institutes and Technology Parks

Applied research should expediently lie in the hands of whom it serves. For Niedersachsen this would mean giving impulses for the establishment of economy related research institutes sponsored by business. Thus, a row of facilities have been founded and sponsored with considerable state funds. These institutes are in each case carried by associations with many, mostly small to medium-sized, businesses as members. The institutes pursue practice-oriented research and development, contract research, and business advisory services as well as training and vocational training.

Regional economic focal points are set with the establishment of the institutes. That is why the state government was also led by regional political considerations in its decisions on the location of the institutes (German Institute for Food Processing Technology in Quakenbrück, Research Institute for Solar Energy in Hameln). That will also be the way in the future, e.g. regarding the decision on the location of the institute for wind power engineering, that is to be located on the coast. These institutes, that owe their foundation to suggestions made by the state as well as the determined interest of the businesses, are sectoral innovation centres and impulse suppliers for big and small entrepreneurs. There should be no doubt that it is the ideal situation when political suggestions and economic interests of businesses also lead to the foundation of regional centres for technology transfer and innovation. I consider the example that "Elf Aquitaine" has given us, with the foundation of the first technology centre in the South west of France in 1979, to be worth imitating. The foundation of CREATI (Centres Régionaux d'Appui Technique et d'Innovation), of regional centres for technical assistance and innovation by the French in 1985 is another step in the commitment of private businesses to regional economic development, in well-understood self-interest, that is worthy of imitating. The "just-in-time concept" e.g. of the motor industry, could lend speed to the foundation of similar private innovation centres in the regions of the Federal Republic of Germany. However, when businesses understand their own task, the advancement of structural change, as an innovative process in a sectoral rather than in a regional dimension, governments have to give an energetic impulse to awaken the interest of the businesses.

Technology policy should at all times only give impulses, but, to my mind, these should be given decidedly when too many businesses still believe they will reach the future unharmed as spectators. An example for impulses for regionally oriented technology policy is i.e. the installation of technology parks or founder centres that, next to the help they provide to their founders, are focal points for the development of a new economic structure at those locations. There is already the result that a new and better innovative climate has come into existence in their vicinity, that reaches beyond the bounds of each town of location. The installation of these founder centres is also worth mentioning because, at least in Niedersachsen, the initiative for their foundation always has come from the region itself. With its decision, the region thus declared an innovation-oriented perspective as important for itself. Except for start–up financing in the form of an investment grant, the Ministry of Economic Affairs has not given any financing assistance for current expenses. After all a regional development strategy, as I have mentioned before, can only be successful, when it is carried out by those responsible. This must also be realized in the cost consciousness of these institutions.

3.2 Technology Transfer

A further example for a regionally oriented technology policy is the installation of innovation consultation agencies at Niedersachsen's Chamber of Industry and Commerce and Chamber of Handicrafts, as well as the technology contact agencies or the technology commissioners at all Niedersachsen's colleges and universities. These "mediator" agencies serve the development and the strengthening of regional initiatives in a special way. In spite of the work of these and other mediators, there are a number of weak points in the co-operation between science and the small and medium-sized firms that are supposed to be ameliorated through the foundation of a permanent alliance of technology mediators and a Niedersachsen agency for technology transfer and innovation promotion (Niedersächsische Agentur für Technologietransfer und Innovationsförderung; NATI). With these two facilities, an increase of efficiency of the existing technology transfer facilities is to be achieved in a way that the information available in these institutions will be improved concerning connection and accessibility and that an additional synergy effect will be achieved through better co-operation.

There are two examples of present technology policy in favour of a region in Niedersachsen that I will mention:

1. The Technology Pool Ostfriesland run as a private limited company that has set the goal to make the best use of the know-how of the Ostfriesland college for regional development in Emden.
2. The conclusion of a contract of co-operation between the rural district of Emden and the Technical University in Braunschweig (Brunswick). The contract includes periodic meetings and information arrangements between the district's businesses and faculties of the Technical University. Thus, information on new technologies is guaranteed and a contact exchange for the regional economy is provided.

4 Information and Communication Policy

I have already mentioned the central role of information and communication techniques under the key word of infrastructure. In general the availability of a modern infrastructure for information and communication and its effective use for macroeconomic efficiency and competitiveness is decisive. Therefore, the innovation and growth potential has to be used carefully. The information and communication techniques and their use will make new products possible and will open new markets. Especially for structurally weak and peripheral regions the new information and communication methods offer a chance to decrease disadvantages of location through early connection to the economic centres. The new methods also serve as a motor for the acceleration of the extension of infrastructure and create the possiblity of a bit more equality of opportunities between the city and the periphery with regard to economic development. The information and communication techniques can bridge distances more easily, but the postal services' policy of charges also has to help to make this possible. Therefore, I am pushing towards a structure of charges to make the advantages of the new information and communication techniques accessible to the structurally weak regions without the disadvantages of higher costs.

I am happy to report that there is a bundle of regional initiatives in Niedersachsen for an offensive use of the information and communication techniques. It should not be seen as a coincidence that these initiatives began in the geographical edges of the state of Niedersachsen, like

1. The Telematics-Centre in Norden (Ostfriesland)
2. The Telehouse Nordhorn (near the border to The Netherlands)
3. The Technology Centre Buxtehude (at the gates of Hamburg)

The field of telematics, which these facilities are dedicated to, is supposed to support the use of new information and communication techniques through technical and organizational advice and training. In spite of the structural problems, especially in the western parts of Niedersachsen, the projects can, and should, contribute to the improvement of the competitiveness of the regional economy by reducing the disadvantages of location and by creating qualified jobs.

In the past, many of the fields mentioned like infrastructure policy, investment promotion policy and technology policy have pursued measures in isolation from each other. In the past, related policies, like ecology and education policy were also pursued independently. In the discussion on economic policy more consideration should be given to the fact that success in structural and economic policy can only come from an interconnectedness of the various political fields.

5 New Factors of Regional Development

Noneconomic basic conditions like the cultural field or the quality of educational establishments are just as important for economic developments as is the condition of the economic structure in a narrow sense: for the economy the question of waste disposal also becomes increasingly a central condition for development. In the future new economic developments will occur first, where the unavoidable and unusable production waste can also be removed or burned. And from this emphasis on ecological factors a new innovative wave of development arises in the economy, with the possibilities of environmental technology that are to be used for regional developments. A test is being made in Niedersachsen to coordinate these different onsets in the interest of an innovation-oriented regional policy, in order to facilitate a form for the future. In "regional conferences" I have promoted the participation of regionally responsible people from businesses, unions, churches, scientific facilities, community administrations and politicians of the region. The first of these regional conferences took place in 1987 in Ostfriesland, a historically structurally weak region. After this conference working groups continued the themes. Next year another conference will take place that will evaluate suggestions for the further development of the region and it is supposed to decide on steps for their realization.

For 1988 a conference for the region of South-Niedersachsen/North-Hessen was suggested by me. Together with the Economic Affairs Minister of the state of Hessen this regional conference was held in Göttingen. A follow up conference is supposed to take place in Kassel, because my colleague from Hessen and I believe that in some cases "regions" within the Federal Republic of Germany have to be defined with overlapping state borders. In the case of South-Niedersachsen/North-Hesse, East-Westphalia being part of the state of North-Rhine-Westphalia should be included as a next step. The regional conferences carried out so far have received great attention. The already existing interest as well as the course and the aftermath of the conferences have proven that apparently the lack of coordinated teamwork between parts of the regions has been strongly felt and that new opportunities for co-operation are being seen. Another task of these regional conferences is to find synergy effects between the various hitherto separated specialized policies. Only in this way can new perspectives be found for structurally weak regions.

Such an increased emphasis on regional self-development does not mean that the Niedersachsen state government should retreat from its responsibility for the regions and their communities. The opposite is true. We are looking for new ways for the realization of a modern strucural policy. One idea being followed up is that of partnerships between businesses, their groups and associations, also including the trade unions as well as the communities and the state. On the other hand, we aim for a definition of common interests that enables the formulation of a joint direction in the co-operation for regional economic development.

The regional conferences have turned out to be forums where representatives of the regional economy, but also representatives of the community, public institutions, groups and associations and politicians can approach each other. Forms of isolations that had lasted for decades between the small regions have been pushed aside. Discussion and agreement on the strategies to be chosen and concrete plans have become the motor for the development of the region. It is significant that the impulses for such regional conferences have been received positively at first in those regions where the economic development was momentarily weakest.

6 Conclusions

So let me point out again: The idea of regional development, for which the regional conferences are only one possible instrument, should no longer be that of trade promotion alone. Structural regional development, in order to make use of the development potentials, has to especially focus on the connection and combination of various special aspects. As I have said before, transportation connections and tourism are just as linked as transportation structures and cultural potential, as economic and educational institutions in a region. Therefore, it is important to define these aspects mutually and to pursue the goals together. It is also important to realize that with all the innovation-oriented regional development strategies the aim should not be, to regard exclusively "high-tech" as "innovation oriented". It is rather the combination of high-tech, low-tech and no-tech aspects that leads to success.

The secret of success lies in a number of various innovation-oriented regional development steps that have to be bundled into a strategy. A regional policy that is oriented in such a way opens new possibilities for action for the regions and states. The primary tool is not to give new financial aids but rather the well-aimed use of existing possibilities and an improved inclusion of the businesses' regional activities. At the end, of course, additional money from the state and federal side and the economy has to be added. But the most important thing is that first of all there is a will in the region to take one's fate into one's own hands and to define what is necessary next. Psychologically it is the same with a region as it is with the individual: motivation grows with the prospect of the success of one's own activities. Without belief in one's own capabilities, the financial help of state and federal governments would not support anything but the mentality of alms giving. The optimum possibility for the realization of the common regional, economic, cultural, ecological and social interests lies in a joint policy of all public bodies and of all those involved on each level. Here lies also the key for the success of a co-ordinated regional development strategy.

The most important and alluring thing about it is that structural policy "from the top" or isolated proceedings of departments are not as accentuated as in the past, but rather the change in the conscience of a region and the permeating idea of its own initiatives. It is no longer the moaning for help from the outside that stands in the foreground. There is rather the demand for support with reference to the region's own accomplishments. The state, in this case the State Government of Niedersachsen, will give further impulses, offer participation and help, provide contacts whereever necessary and also grant financial support for the realization of a regionally developed overall concept.

The conclusion of all I have said is: We have to succeed in defining the region- and location-specific features and the possibilities for their development. This must be done within the regions and by the regionally responsible themselves. But the economic policy cannot be successful when the conditions in related fields like culture, education and the environment are not included in the planning. Also success will only come if the structural change in the economy is not hindered and innovations are supported consequently in all fields of the economy. This also means that it can not as yet be conclusively defined where any single development is going to go. It is important to keep an open eye for the new and to win perspectives for regional development so that innovation is not only the further development of the known but rather a basic openness for new ways and methods. Innovation orientation as a role model for successful regional development strategies, therefore, means not only having an eye on the situation and the possibilities for the development of the economy, but also on the living conditions in a region as a whole, however, the "region" might be limited. There cannot be a general priority for the regional. But the motto "in dubio pro regio" should more and more become a touchstone. The necessary decisions of the central state, e.g. for transportation councils, for national large-scale research institutes or for supraregional abatement facilities are one thing. The right tuning between these measures and the regional encouragement and the seizing of regional initiatives, whether we call them "domestic forces" or "endogenous potential", is the other, equally important thing. If everyone actively perceives his role, this will be to the benefit of the central government and the regions.

6.2 Regional Economic Development and Innovation-Oriented Measures: Summary and Perspectives

Hans-Jürgen Ewers

In light of the multiplicity of themes and papers presented, the task of presenting a summary of the most important results of this volume is an assignment which will perforce to draw the anger of several of the authors for having omitted their ideas or not having given those ideas the consideration which they would have desired. In order to be in equally hot water with all contributors I shall not make reference to any one paper. Instead, I will try to answer briefly five questions which in my opinion have been part of the centerpiece of this volume and for which most of the authorshave tried to provide answers:

1. Which regional development trends will affect the economy in the coming years?
2. What forces are behind these trends?
3. Why do revitalization strategies need to be conceived and carried out above all at the local level?
4. What things are important to regard when attempting an endogenous, innovation-oriented strategy?
5. Which questions remain open?

1 Regional Development Trends

In light of the variety of regional developments within the countries of Europe taking place during the 1970s and 1980s, it seems nearly impossible to make a general statement concerning regional development trends. If such a statement were to be attempted, then it would have to include the supposition that the decentralization of economic activities (as opposed to that which we were accustomed to up to the beginning of the 1970s) will most likely continue through the 1990s. Decentralization of economic activities is characterized by a negative correlation between the status of a location within the hierarchy of cities and the growth rate of its employment in industry as well as in the service sector. Seen as a group, the large agglomerates of employment have suffered setback, while rural regions have gained either in relative or absolute employment.

Although the more or less intensely perceived reversal within the countries of Europe ever since the beginning of the 1970s from the previously dominant trend toward global regional concentration will probably be of long-lasting nature, this will not mean that every agglomeration will be consigned to reduction, nor that every rural location will encounter good opportunities for growth. The variance in development tendencies between locations of equal status has increased drastically since the beginning of the 1970s. In other words: For any regional class we can find unusually prosperous as well as radically diminishing regions. The southerly movement of growth in the Federal Republic of Germany, for example, is one manifestation of this increased developmental variance affecting locations at all levels of municipal hierarchy.

2 What Forces are Behind these Trends?

Therefore, if one wishes to assess the possibilities of a location with regard to these general developmental trends so as to be able to build a developmental strategy, the forces behind these trends will have to be analyzed. This is where I would like to focus attention on two particular types of forces:

Forces influencing the general (sectorial) structural transition and

Forces which are crucial for the particular regional impact of this general structural transition.

The general structural transition will be affected in the coming years above all by three factors which have also been decisive for the direction of structural transition occurring in the 1980s:

1. A first major factor stems from the now-as-ever changing international division of labour. The "new international division of labour" by which ever more former developing countries become producers of industrial goods for a world market has already introduced all those producers in the old industrial nations to pressure from competition which try to combat with low prices and by inputting less qualified workers. Such standardized mass production activities will not be able to survive long-term in the high-wage

countries of Europe, even if its departure may possibly be protractable for some period of time by increasing the strategy of automation. In the long-term, technologically complex, creative production activities as well as services oriented toward these activities' needs will have to support the growth of high-wage countries.

2. At least an equally strong influence is attributable to the new shortage of resources which most western industrial nations have been subject to since the beginning of the 1970s. This shortage of resources concerns all raw materials, particularly prime energy carriers and also the natural basics for living which have long since become scarce in the old industrial nations. Accordingly, the structural transition now induced by the new shortage of resources will affect above all those economic activities continuing to rely on raw materials and environment for production, i. e., for the most part the "old" industries. Human-capital reliant as well as clean industries are advantaged by this structural transition; these particularly include production-oriented services which can be exported (e.g., as results of research & development, design, etc.) throughout the world.
3. A third influence on structural transition which will probably gain ground over the 1990s are the new technologies related to the computer and miniaturization of components. These new technologies were applied during the 1970s and 1980s to the rationalization of old production and, as such, served to underscore the established trends. Increasingly, these technologies will be used for product innovation and will create their own, individual developmental patterns. The best-known example of such an individual pattern can more recently be seen in the trend toward "systemic rationalization" which - in contrast to such task-specific rationalization as was previously common - is not restricted to single plant/operation functions, but rather has long-since sprung plant boundaries to optimize entire production chains including outside suppliers (as in "just-in-time" concepts). Systemic rationalization means the reorganization and integration of all sub-processes carried out at a plant as well as at that plant's suppliers using digital communication networks and computer control to gain greater flexibility via "soft" connecting of machinery and to reduce production costs particularly by lowering circulating capital. Flexibility on the one hand versus production-cost reduction on the other - two contradicting aims under conventional automation - can now be achieved simultaneously using computer control and telecommuni-

cation. Systemic rationalization will considerably affect the regional production structures. This is because on the one hand the possibilities increase for decentralization of physical production while on the other hand possibilities become greater for centralizing regulation and control activities.

Just how these global developmental trends will manifest themselves regionally, i.e., which regions will gain and which will lose, depends on the characteristics of the specific structure of each individual region. The analyses presented in this volume have shown that in particular the following characteristics have been crucial for successful regional growth:

1. the sectorial structure of the region (particularly the sectors' ages, the intensity of human capital, and the degree of standardization of the products produced),
2. the size structure of the plants in the region (large-operation structured regions suffered much more heavily from the structural transition than did smaller- operation structured regions),
3. the functional structure of a region (the mixture of manufacturing and service activities as well as - among the service activities - the mixture of creative and routine professions),
4. the location's hierarchic position within the spatial system (because there are typical differences between agglomeration cores and periphery with regard to the age of the economic sectores, the qualification structure of the working input, and the extent of product or process innovation).

3 Why do Revitalization Strategies need to be Conceived and Carried out Above all at the Local Level?

The majority of papers presented in this volume assume either explicitly or implicitly that - in comparison to the industrial policies of the 1960s and 1970s - today and in future, the local (regional) level will receive considerably more attention during the conception and implementation of revitalization policies. Four groups of reasons can be named for the increased attention to the local level in implementing revitalization strategies:

A first reason is the different degree of effect which each region is subject to as described above from the structural transition in relation to their specific characteristics. Accordingly, not only different revitalization concepts are needed for different regional classes but one must also differentiate within a particular class of regions the considerably varying degrees of effect each region is subject to and take these into account for the political concept required.

Apart from the different effect upon each region, considerable interregional differences with respect to basic factors can be detected. Such differences manifest themselves in:

1. Infrastructural provisions (R & D institutions, available service institutions, communications infrastructure),
2. Environmental situation within the region (degree to which the natural basic living conditions are being used), and in the
3. Multiplicity and performance of political networks, or the formal and informal relations between private, intermediary, and government administrative agents and agencies in the region.

An additional reason for an increased significance of the local level in revitalization strategies can be seen in the change of the addressees that regional economic promotion has initiated over the past years: no longer does the attempt to settle large firms or to retain jobs in big plants stand in the center of region-promotional policies, rather, the founding of new firms and promotion of growth opportunities for small and medium-sized enterprises (SMEs). This change is not so much the result of a superficial "pick the winner" strategy following on a noticeable rise in net job openings within the group of SMEs as much as it reflects an understanding off the particular resource deficits of SMEs in the process of innovation. Most of the analyses conducted in this area show SMEs as having considerable deficits with regard to important innovation-relevant resources. This refers in particular to employee qualification, technological know-how, and access to information networks. Because SMEs are generally restricted in their access to these resources to the narrow confines of their immediate surroundings, any attempt to compensate such deficits need to take place within the same narrow regional field. The majority of SME businessmen and -women depend now as ever on just such help-networks which they know and trust within their immediate local area.

A last reason which forces revitalization strategy activity onto the local level is also the change in recent years of instruments used in regional economic promotion: justifiably, economic sponsors are trusting less in financial stimulus instruments since these create considerable free-loader effects and are not efficient, often missing the real problem or deficit within the firm and weighing heavily in favor of larger company operations. Instead, more weight is being placed on real transfers geared to the needs of local firms. Real transfer stands for the offering of all types of public services, from consulting and counselling, to qualification measures, to active help when applying for promotion measures.

4 What Things are Important to Regard when Attempting an Endogenous, Innovation-Oriented Strategy?

I would like to attempt a formulation of the basic principles important to the conception and implementation of an endogenous, innovation-oriented strategy by way of a catalogue of ten "commandments" of regional economic promotion.

1. Do nothing before having made a thorough analysis of the economic situation in your region. Important knowledge is that of sectors and products, technologies and qualifications, firms and large plant operations, and the extent of external control, infrastructure, and institutional structure of the region.

2. Do not try to become manager of the short-term crisis. Any attempt to retain jobs which have been made obsolete by structural transitions will almost always lead to wasteful use of promotion funds which could be considerably more effectively invested over long term in the creation of new jobs.

3. Do not follow the high-tech apostles. Promotion strategies directed toward high tech demand infrastructural preconditions (as in R&D) available only in very few large cities.

4. Do not confuse business promotion with technology promotion. The creation of a reliable management information system is often more important for the development of a local business than owning the latest computer-controlled equipment.

5. Do not sit in your promotional chair and wait for a somebody to come with a request for innovation strategy. The problem of most stagnating businesses lies in the fact that they are not convinced by the necessity of a strategy for innovation.

6. Do not wait to contact a firm until the development crisis has arrived. Crisis situations must be anticipated and countered by measures within a framework of long-term, predicting strategy.

7. Do not rely - exclusively - on your own ability as a consultant. No regional promoter of business can be a specialist for everything. The promoter's task should be to find out which expert is needed as an advisor.

8. Do not trust solely in the consultant. Even when an advisor has been selected with painstaking care, the problem often arises that a firm and the consultant are not suited to

one another or that the firm is not able to carry through with the consultant's recommendations. The promoter should assume the role of a moderator, and his observation should accompany the implementation of an innovation strategy in the firm he serves.

9. Do not try to create a large promotional institution for local enterprise. The task of the promoter is not to create the needed jobs within a local business-promotion agency itself.

10. Do not trust in what your own staff tells you about the successful performance of your promotion institution. Outside evaluation belongs to the scope of any professional promotion strategy and usually provides the breeding ground for important impulses of change.

5 Some Unsettled Questions

Just as with every book, this one, too, has posed several new questions and directed attention in the discussions toward others older and heretofore unanswered. I would like to point out four of these questions which could, each one for itself, provide the opening theme for a new volume.

1. If one accepts that the building of networks between private, intermediary, and public agencies of a region and the creation of an innovative business climate in the region are important - if not the most important - tasks of a promoter of regional enterprise, then the question arises as to which network design is the most efficient: is it better for a region's economic development to be faced with a sleekly centralized, hierarchical network (a "one-stop" agency), or had there better be competition between different public business-promotion institutions or between public as well as private agencies?

2. What are the possibilities for privatizing the business of promoting business? Do reasons exist which could lead to the inevitable failure of a "Market for the Promotion of the Economy"? What prerequisites must be met before a private marketplace for the required advisory, qualification, and transfer services can be established?

3. What qualifications should public promoters of enterprise possess, and how should they be paid? Can public employees - remunerated according to the customary practice of the civil service - even be used to carry out successful business promotion? What possibilities exist for success-oriented remuneration of regional business promoters?

4. What is the nature of relations between small and large firms within the process of technical transition and the dynamics of employment? Are those economy promotion policies which are exclusively directed toward SMEs perhaps incomplete because they disregard the essentiality of symbiosis of large and small businesses? In what ways can the existing large firms of a region be incorporated in a revitalization strategy for the benefit of smaller industry?

Contributors

Kevin Allen	Professor and Director of the European Policy Research Centre at the University of Strathclyde, U.K.
Jürgen Allesch	Managing Director of the Technologie-Vermittlungs-Agentur Berlin e.V.
Franz-Josef Bade	Professor of Economics at the University of Dortmund
Ernst A. Brugger	Privatdozent at the University of Zürich and Economic Consultant
Roberto Camagni	Professor at the Universitá Commerciale Bocconi, Milano
Evert Jan Davelaar	Institute of Economics of the Free University of Amsterdam
Hans-Jürgen Ewers	Professor of Economics at the Technical University of Berlin
Jürgen Gabriel	Institut of Political Science of the Free University of Berlin
Jean-Paul de Gaudemar	Professor of Economics at the University of Aix-Marseille and President of the OECD Working Party on Regional Development Policies, Paris
Joachim Genosko	Professor of Economics at the University of Hohenheim
Ottokar Hahn	Minister for Federal Affairs and Special Tasks of the Saarland, Saarbrücken
Walter Hirche	Minister for Economic Affairs, Technology and Transport of Niedersachsen, Hannover

Benny Hjern	Professor of Political Sciences at the University of Umeå, Sweden
Christopher Hull	Assistent Secretary General of the European Association for the Transfer of Technology, Innovation and Industrial Information (TII), Luxembourg
Gwyn Jones	President of the Welsh Development Agency, Cardiff
Manfred Kern	Director of the Department for Economics of the Chamber of Industry and Commerce at Berlin
Edward J. Malecki	Professor of Geography at the University of Florida, Gainsville, USA
T. Joseph Mulcahy	Directorate General XVI (Regional Policy) of the Commission of the European Communities, Brussels
Harry Nicholls	Managing Director Aston Science Park, Birmingham U.K.
Peter Nijkamp	Professor of Economics at the Free University of Amsterdam
Kathrin Peters	Segal-Quince-Wicksteed, Economic Consultants, Cambridge, U.K.
Elmar Pieroth	Former Senator for Economics and Labour of Berlin
Roberta Rabellotti	Institute of Economics of the Università Commerciale Bocconi, Milano
Gerry Sweeney	Director SICA Innovation Consultants, Dublin
Peter Townroe	Professor of Economics at the University of East Anglia, Norwich, U.K.
Douglas Yuill	European Policy Research Centre at the University of Strathclyde, UK

Bei Fragen zur Produktsicherheit wenden Sie sich bitte an:
If you have any questions regarding product safety,
please contact:

Walter de Gruyter GmbH
Genthiner Straße 13
10785 Berlin
productsafety@degruyterbrill.com